TEACHER LEARNING OF AMBITIOUS AND EQUITABLE MATHEMATICS INSTRUCTION

Drawing on sociocultural learning theory, this book offers a groundbreaking theory of secondary mathematics teacher learning in schools, focusing on the transformation of instruction as a conceptual change project to achieve ambitious and equitable mathematics teaching.

Despite decades of research showing the importance of ambitious and equitable teaching, few inroads have been made in most U.S. classrooms, and teacher learning in general remains undertheorized in most educational research. Illustrating their theory through closely documented case studies of secondary mathematics teachers' learning and instructional practices, authors Horn and Garner explore the key conceptual issues teachers are required to work through in order to more fully realize ambitious and equitable teaching in their classrooms. By theorizing teacher learning from a sociocultural perspective and focusing on instructional practice, the authors make a unique contribution to the field of teacher learning.

This book offers researchers, scholars, and teacher educators new theoretical and methodological tools for the elusive phenomenon of teacher learning, and provides instructional leaders and coaches with practical examples of how teachers shift their thinking and practice.

Ilana Horn is Professor of Mathematics Education at Vanderbilt University Peabody College, USA, where she heads the Teacher Learning Lab.

Brette Garner is Assistant Professor of Mathematics Education at the Morgridge College of Education at the University of Denver, USA.

Studies in Mathematical Thinking and Learning
Alan H. Schoenfeld, Series Editor

TEACHER LEARNING OF AMBITIOUS AND EQUITABLE MATHEMATICS INSTRUCTION

A Sociocultural Approach

Ilana Horn and Brette Garner

Routledge
Taylor & Francis Group

NEW YORK AND LONDON

Cover image: © Getty Images

First published 2022
by Routledge
605 Third Avenue, New York, NY 10158

and by Routledge
2 Park Square, Milton Park, Abingdon, Oxon, OX14 4RN

Routledge is an imprint of the Taylor & Francis Group, an informa business

Library of Congress Cataloging-in-Publication Data

Names: Horn, Ilana Seidel, author.
Title: Teacher learning of ambitious and equitable mathematics
    instruction: a sociocultural approach/Ilana Horn and Brette Garner.
Description: New York, NY: Routledge, 2022. | Series: Studies in
    mathematical thinking and learning | Includes bibliographical
    references and index.
Identifiers: LCCN 2021048105 | ISBN 9781032021768 (hardback) |
    ISBN 9781032021744 (paperback) | ISBN 9781003182214 (ebook)
Subjects: LCSH: Mathematics—Study and teaching. | Learning—Social aspects.
Classification: LCC QA11.2.H655 2022 | DDC 510.71/2—dc23/eng/20211216
    LC record available at https://lccn.loc.gov/2021048105

ISBN: 9781032021768 (hbk)
ISBN: 9781032021744 (pbk)
ISBN: 9781003182214 (ebk)

DOI: 10.4324/9781003182214

Typeset in Bembo
by KnowledgeWorks Global Ltd.

To the teachers who generously shared their classrooms, their thinking, and their commitment to their students with our research team. This truly would not have been possible without you.

CONTENTS

FIGURES

TABLES

ABOUT THE AUTHORS

Ilana Seidel Horn is a Professor of Mathematics Education at Vanderbilt University Peabody College, Nashville, TN, where she heads the Teacher Learning Lab. Professor Horn's work lies at the intersection of several disciplines: mathematics education, the sociology of schooling, and learning sciences. She uses sociolinguistics and interpretive methods to examine secondary mathematics teachers' learning in the contexts of their workplace, yielding images of teachers' learning and practice that account for the institutional setting of schools and the pressures of policy. By understanding teachers' learning in the workplace, her research highlights ways to make teacher education and professional development usable and relevant to practicing educators.

Brette Garner is an Assistant Professor at the Morgridge College of Education at the University of Denver. Her research focuses on how mathematics teachers learn ambitious and equitable instruction, particularly in light of the sociopolitical forces that constrain their work. As a teacher educator and researcher, Dr. Garner works with pre-service and in-service teachers to cultivate more equitable, rehumanizing, and engaging mathematics classrooms, where students can grow as competent and capable mathematical thinkers.

PREFACE

Teacher learning is central to many educational change projects. Whether policymakers seek to implement new instructional standards, parent advocates want greater and more meaningful inclusion, or researchers hope to promote certain practices to support student learning, these efforts rest on classroom teachers' capacity to incorporate new practices into their instruction – to learn to teach differently. Despite its centrality to the educational enterprise, teacher learning remains undertheorized in most educational research, including research that focuses on teacher education and professional development.

Over the years, we have pursued questions of teacher learning in the context of secondary mathematics teachers who are transforming their practice to reflect the values of ambitious and equitable instruction. Ambitious mathematics teaching aims to support students' development of both conceptual understanding and procedural fluency; ambitious and *equitable* teaching underscores the need to support such learning in *all* students, with particular attention to those who have been historically underserved in schools (Cobb et al., 2018).

In some of our studies, the impetus for the move toward ambitious and equitable instruction comes from the teachers themselves (e.g., Horn, 2005, 2010). In others, it comes from external sources, such as school districts (e.g., Horn et al., 2017, 2020). We have studied this learning in the context of teacher education (e.g., Horn et al., 2008; Horn & Campbell, 2015), through accountability policies (e.g., Garner et al., 2017; Horn, 2018; Horn et al., 2015), in teacher workgroups (e.g., Horn & Kane, 2015; Horn & Little, 2010), and in professional development (Chen et al., 2020; Horn, 2019). Across these studies, we approach questions of teacher learning with an ethnomethodological lens, emphasizing the meaning teachers are making from these various activities (Garfinkel, 1967).

Inevitably, our focus on teachers' reasoning about their work as they shift their practice uncovers the institutional logics and obstacles they encounter, which often impede their goals. For this reason, we see profound ecological validity in reflecting these contextual realities in our analysis of teacher learning. Furthermore, teaching, as something to "know," can only be fully realized in interaction with others; we see (and have ourselves experienced) teachers who succeeded with a particular practice in one setting struggle to have the same results elsewhere. Teachers are not super-agentic beings who can bypass the everyday realities of their work; it is therefore, in our opinion, insufficient to study teacher learning only as changes in individual cognition. Instead, sociocultural theories of learning offer lenses to recognize how notions about teaching do not merely reflect what individual teachers think, understand, and do. Rather, these notions also reflect the cultural norms and practices of schooling itself.

Sociocultural perspectives stem from Soviet psychologist's Lev Vygotsky's original insights about the role of language, cultural artifacts, and interaction in human development. Although his ideas have been taken up by various scholars – from psychologists (e.g., Rogoff, 2003; Wertsch, 1985) to anthropologists (e.g., Holland & Valsiner, 1988; Lave, 1988) to critical theorists (e.g., Esmonde & Booker, 2016; Nasir & Hand, 2006) – we position ourselves on the anthropological tradition of this spectrum. Across these approaches, analysts shift their lens from questions of changes in *individuals over time* to *individuals in contexts over time*. What varies is the depth to which those contexts are envisioned – the immediate context of interaction, cultural norms and practices, or flows of power. Inevitably, looking at institutional influence on teacher understandings uncovers flows of power, so we account for it where we find it, but these perspectives are not as central to our work on teacher learning as they are in other scholarship (e.g., McKinney de Royston et al., 2021) and we aim to be transparent about that.

The subfield of sociocultural learning theory we use is best captured as what James Greeno (1998) called a *situative perspective*. Some primary tenets of this perspective are that: (1) information is organized in analyzable ways in any socially organized activity; (2) people adopt, transform, and resist this information in ways that can be captured through interactional analysis; and (3) socially organized activities have characteristic practices, which people similarly adopt, transform, and resist, and these practices convey meaning. In our situative studies of teacher learning, we take schooling in general – and mathematics teaching in particular – as the socially organized activity of interest; the meaning systems that we focus on are those that convey messages about what teaching is, what mathematics is, what learning is, and who students are. Teachers, as they encounter new ideas, negotiate these messages with other meaning systems; context is thus not additive to an analysis of their learning but a fundamental facet of it.

Looking at how meanings develop within broader systems, sociocultural lenses also invite us to interrogate different dimensions of and timescales for teachers' learning. Because meanings about teaching, learners, and mathematics do not simply "exist," this perspective raises questions about how meanings came about – the sociohistorical conditions of their making. This happens on multiple timescales: Meanings about these core ideas are reinstantiated, resisted, and negotiated in the microculture of teacher communities (Horn, 2007) and in broader social groups, like the parent communities that mobilized in the U.S. Math Wars (Schoenfeld, 2004). Looking at meanings' sources across timescales and communities, in turn, invites analyses of the tensions that teacher learners encounter as they make sense of their actions in various social worlds (Holland et al., 2001) – schools, communities, and the broader educational landscape. Research on teaching is already rich with the vocabulary of dilemmas (e.g., Lampert, 1985); a sociocultural perspective offers a way to incorporate such insights into a theory of teacher learning, tracing them back to their institutional (Horn, 2018), historical (Schneider, 2014), and ideological (Philip et al., 2016) origins.

This monograph summarizes our understanding of secondary mathematics teachers' learning about ambitious and equitable instruction from a situative perspective. In Part I, we synthesize prior work on sociocultural learning theory, with an emphasis on what it illuminates about teachers' learning. We read systematically within mathematics education, sociocultural studies of teacher learning, and sociological studies of teachers' work, drawing eclectically from a range of sociocultural theorists whose ideas influence our thinking. Our central argument is that teacher learning can productively be viewed as concept development, with *narrativized actions and understandings* as a way to trace shifts that signal learning. In Part II, we describe how we used our theoretical lens to design an intervention to facilitate the learning of 12 highly committed urban secondary mathematics teachers, members of a professional development organization (PDO) whose purpose was to support teachers toward this goal. Our research study – *Supporting Instructional Growth in Mathematics*, or *Project SIGMa* – took place over four years as a research-practice partnership with the PDO. Our partnership responded to a felt need from the teachers and the PDO leaders – teachers needed more feedback on their instruction to support their learning. This led to our design of a *video formative feedback* (VFF) cycle, a video-based coaching model that built on teachers' questions. Our findings detail how our intervention supported our participants' development. Using these cases, we focus on teacher learning as refining *pedagogical judgment*, a dialectical process between teachers' action and reasoning given their responsibilities, arguing that this is central to how practice changes.

Our intended audience for this monograph is wide in scope. Clearly, our primary audience is mathematics educators who share our goals – more ambitious and equitable mathematics instruction in schools – including teacher leaders, teacher educators, instructional coaches, and professional development providers. Additionally, we hope that our theoretical frameworks contribute to broader research in pre-service and in-service teacher education, where teacher learning is often undertheorized. Finally, because many of the impediments our participants encountered in enacting ambitious and equitable instruction stemmed from the grammar of schooling itself (Tyack & Tobin, 1994), our most aspirational wish is that our study encourages changes in teachers' work conditions (which are, of course, also their learning conditions). This is undoubtedly needed if we, as a society, are to take the goal of ambitious and equitable instruction seriously.

References

Chen, G. A., Marshall, S. A., & Horn, I. S. (2020). 'How do I choose?': Mathematics teachers' sensemaking about pedagogical responsibility. *Pedagogy, Culture & Society, 29*(3), 1–18.

Cobb, P., Jackson, K., Henrick, E., & Smith, T. M. (2018). *Systems for instructional improvement: Creating coherence from the classroom to the district office.* Harvard Education Press.

Esmonde, I., & Booker, A. N. (Eds.). (2016). *Power and privilege in the learning sciences: Critical and sociocultural theories of learning.* Taylor & Francis.

Garfinkel, H. (1967). *Studies in ethnomethodology.* Paradigm Publishers.

Garner, B., Thorne, J. K., & Horn, I. S. (2017). Teachers interpreting data for instructional decisions: Where does equity come in? *Journal of Educational Administration, 55*(4), 407–426.

Greeno, J. G. (1998). The situativity of knowing, learning, and research. *American Psychologist, 53*(1), 5–26.

Holland, D., Lachicotte Jr, W., Skinner, D., & Cain, C. (2001). *Identity and agency in cultural worlds.* Harvard University Press.

Holland, D. C., & Valsiner, J. (1988). Cognition, symbols, and Vygotsky's developmental psychology. *Ethos, 16*(3), 247–272.

Horn, I. S. (2005). Learning on the job: A situated account of teacher learning in high school mathematics departments. *Cognition and Instruction, 23*(2), 207–236.

Horn, I. S. (2007). Fast kids, slow kids, lazy kids: Framing the mismatch problem in mathematics teachers' conversations. *The Journal of the Learning Sciences, 16*(1), 37–79.

Horn, I.S. (2010). Teaching replays, teaching rehearsals, and re-visions of practice: Learning from colleagues in a mathematics teacher community. *Teachers College Record, 112*(1), 225–259.

Horn, I. S. (2018). Accountability as a design for teacher learning: Sensemaking about mathematics and equity in the NCLB era. *Urban Education, 53*(3), 382–408.

Horn, I. S. (2019). Supporting the development of pedagogical judgment: Connecting instruction to contexts through classroom video with experienced mathematics teachers. In *International Handbook of Mathematics Teacher Education: Volume 3* (pp. 321–342). Brill Sense.

Horn, I. S., & Campbell, S. S. (2015). Developing pedagogical judgment in novice teachers: Mediated field experience as a pedagogy for teacher education. *Pedagogies: An International Journal, 10*(2), 149–176.

Horn, I., Garner, B., Chen, I. C., & Frank, K. A. (2020). Seeing colleagues as learning resources: The influence of mathematics teacher meetings on advice-seeking social networks. *AERA Open, 6*(2), 2332858420914898.

Horn, I. S., Garner, B., Kane, B. D., & Brasel, J. (2017). A taxonomy of instructional learning opportunities in teachers' workgroup conversations. *Journal of Teacher Education, 68*(1), 41–54.

Horn, I. S., & Kane, B. D. (2015). Opportunities for professional learning in mathematics teacher workgroup conversations: Relationships to instructional expertise. *Journal of the Learning Sciences, 24*(3), 373–418.

Horn, I. S., Kane, B. D., & Wilson, J. (2015). Making sense of student performance data: Data use logics and mathematics teachers' learning opportunities. *American Educational Research Journal, 52*(2), 208–242.

Horn, I. S., & Little, J. W. (2010). Attending to problems of practice: Routines and resources for professional learning in teachers' workplace interactions. *American Educational Research Journal, 47*(1), 181–217.

Horn, I. S., Nolen, S. B., Ward, C., & Campbell, S. S. (2008). Developing practices in multiple worlds: The role of identity in learning to teach. *Teacher Education Quarterly, 35*(3), 61–72.

Lampert, M. (1985). How do teachers manage to teach? Perspectives on problems in practice. *Harvard Educational Review, 55*(2), 178–195.

Lave, J. (1988). *Cognition in practice: Mind, mathematics and culture in everyday life.* Cambridge University Press.

McKinney de Royston, M., Madkins, T. C., Givens, J. R., & Nasir, N. I. S. (2021). "I'm a teacher, I'm gonna always protect you": Understanding black educators' protection of Black Children. *American Educational Research Journal, 58*(1), 68–106.

Nasir, N. I. S., & Hand, V. M. (2006). Exploring sociocultural perspectives on race, culture, and learning. *Review of Educational Research, 76*(4), 449–475.

Philip, T. M., Olivares-Pasillas, M. C., & Rocha, J. (2016). Becoming racially literate about data and data-literate about race: Data visualizations in the classroom as a site of racial-ideological micro-contestations. *Cognition and Instruction, 34*(4), 361–388.

Rogoff, B. (2003). *The cultural nature of human development.* Oxford University Press.

Schoenfeld, A. H. (2004). The math wars. *Educational Policy, 18*(1), 253–286.

Schneider, J. (2014). *From the Ivory Tower to the schoolhouse: How scholarship becomes common knowledge in education.* Harvard Education Press.

Tyack, D., & Tobin, W. (1994). The "grammar" of schooling: Why has it been so hard to change? *American Educational Research Journal, 31*(3), 453–479.

Wertsch, J. V. (1985). *Vygotsky and the social formation of mind.* Harvard University Press.

ACKNOWLEDGEMENTS

We are grateful to all the people who helped us with this study, our analysis, and ultimately the writing of this book. First, we cannot say enough about the wonderful Project SIGMa research team. In order of appearance on the project, they are Grace Chen, Sammie Marshall, Patty Buenrostro, Nadav Ehrenfeld, Katherine Schneeberger McGugan, and Lizi Metts. Together, they contributed thousands of hours of fieldwork, analysis, and writing, and the team worked with intelligence, good humor, mutual support, excellent memes, the occasional prank, and copious snacks. Learning really is better when working in community, and their collaboration and spirit showed us that once again. We know that there are conversations with them that are captured in what seem to be our words, and we thank them for their generous colleagueship. Of course, their contributions to writing this book have also directly shaped this work and our thinking about teacher learning.

We are also grateful to the SIGMa babies whose cameo appearances on our Zoom always took the spotlight. Karmi, Bo, Vivian, Alder, Ali, and James — we are hoping to build a better future for you, too. We are indebted to the SIGMa kids, Xiomara and Ximena, for introducing us to Pusheen the Cat and creating the SIGMa TikTok account. The emotional support and mischievous antics of the SIGMa pets also made our lives better. Tesla, Magnet, Shosh, Affie, Bartleby, Zooey, Rooney, Swizly, and Tiko: thank you for the support and affection.

Additionally, we depended on the extraordinary research support of Mariah Harmon, Lara Jasien, Jess Moses, Maria Aguilera, Yasmin Aguilon, Kathy Forsthoff, Alexandra Lee, Katy Janik, Chi Xiao, Natalie Boyd, Natalie Wyatt, Rebecca Eddy, Ashley Alchehayed, and Emma Reimers. Special thanks to Lee Druce for assistance with swashbuckling and graphic design.

Qualitative research is always an adventure, and our study is stronger thanks to our phenomenal advisory board, who helped us figure out which adventures to choose. Our advisory board members were Hilda Borko, Susan Jurow, Judith Warren Little, Nicole Louie, Chuck Munter, Susan Nolen, Thomas Philip, Miriam Sherin, and Irene Yoon. Extra thanks to our manuscript readers, Hilda Borko, Rebecca Eddy, Susan Jurow, Judith Warren Little, Susan Nolen, Miriam Sherin, Barb Stengel, and Darryl Yong. Their smart, incisive feedback sharpened our revisions. We could not have kept on our writing schedule without the additional assistance from Lizi Metts, who worked tirelessly to consolidate this feedback, turning gritty sand into beautiful pearls from which we could glean all the wisdom, and who probably now knows more about formatting transcripts, APA 7 headers, and tracking document versions than she ever hoped to learn. Additionally, I (Lani Horn) want to thank Elinor Horn for entering my old-school, handwritten edits into word processing documents; her assistance was invaluable in the final phases of writing and revising, and her occasional commentary, charming doodles, and loving encouragement meant the world to me.

On a personal front, we both want to thank our families and partners. Lani is grateful for the support, nourishment, and humor provided by Adam, Naomi, Elinor, and Judah. I can't imagine a more loving group of supporters for Mom's "bookie bookie" project. Brette is thankful for Hope's unwavering love and encouragement, as well as the support of her family – Linda, Brad, Brooke, Brie, and Gary. I am so lucky to have a bright constellation of loved ones guiding my way.

We also honor the memory of our NSF Program Officer, the late Karen D. King, who saw the potential of Project SIGMa and generously shared her wide-ranging insights along the way. The world lost her too soon, and she is truly missed. We are grateful to Margret Hjalmarson for taking on the daunting task of stepping into Karen's unfillable shoes. Margret has proven to be a formidable ally in her own right, and her continued support, guidance, and cheerleading have been tremendous.

Of course, none of this work would be possible without the amazing effort and generous collaboration of the PDO leaders and staff. They have built a wonderful community of thoughtful, dedicated, and accomplished educators; we are grateful that they let us be a part of it.

And most of all, we want to thank the awesome teachers and students who welcomed us – and our cameras – into their spaces and made themselves vulnerable in the process. When we asked for volunteers to let us record their practice, the SIGMa teachers enthusiastically and generously invited us into their classrooms, conversations, happy hours, and celebrations. It is a joy to visit such happy classes, where teachers and students can laugh and joke while learning mathematics meaningfully. And informal conversations with SIGMa teachers – deeply philosophical, nerdily mathematical, wonderfully silly, and everything in between – have been a true delight. We are honored to have had an opportunity to learn alongside them. They have made us better teachers and teacher educators, while showing us a more hopeful vision for the future of education.

Officially, we have to say: This material is based upon work supported by the National Science Foundation under Grant #DRL-1620920. Any opinions, findings, and conclusions or recommendations expressed in this material are those of the authors and do not necessarily reflect the views of the National Science Foundation or other collaborators. But what we really mean is that we are grateful to the many sources of support we have been fortunate to have, and all remaining errors are our own.

PART 1

Theorizing Mathematics Teachers' Learning

As mathematics education researchers, we devote our careers to identifying humane forms of instruction that engage students with the wonder and delight of mathematics. Numerous studies demonstrate that meaningful contexts offer students ways to actively make sense of ideas (Brousseau, 2006; Freudenthal, 2012; Gutstein, 2006; Lampert, 2001). These well-developed examples offer compelling images of student-driven mathematical insight, including how constructivist thinking and positive mathematical identities can transform mathematical learning from the bane of children's schooling to something they genuinely enjoy (Gresalfi & Hand, 2019; Sinclair, 2004).

Simultaneously, our field's lack of clarity around supporting teachers in implementing these forms of instruction hinders the broader enterprise. Certainly, these difficulties do not stem from lack of effort. Researchers have examined questions of teachers' use of ambitious and equitable mathematics instruction through numerous lenses, studying implementation, school reform, curricular design, and so forth. Yet in many cases, teachers' uptake of ambitious and equitable instruction has been limited. From our perspective, we see this difficulty as arising from a significant theoretical gap, since transforming instruction is ultimately a question of teacher learning.

Educational research offers strong theories of student learning, particularly in well-studied domains such as literacy, elementary mathematics, and scientific reasoning. However, we do not yet have well-developed ideas about teacher learning. Instead, teacher learning is typically viewed as an outcome of educational experiences, whether professional development, preservice education, or researcher-designed interventions. These narrow sites of investigation have resulted in thin theory around teacher learning – the details of learning processes often remain unstated, with implicit theories

DOI: 10.4324/9781003182214-1

that educational experiences cause changes in knowledge and beliefs that, in turn, change instruction (Clarke & Hollingsworth, 2002). Missing from these accounts are a sense of how these changes came about.

A central goal of this book is therefore to "thicken up theory" on teacher learning. To do so, we start by taking a conceptual change perspective. Conceptual change, as a framework, moves beyond simplistic forms of learning – like skill acquisition or additive changes to practice – and instead emphasizes the transformation of fundamental understandings (diSessa, 2006). By emphasizing that learners actively construct new ideas, conceptual change shows how processes of learning are influenced by both the learner and the context in which they learn. This perspective has implications for educational designs; effective designs must attend to learners' prior notions – and, we will argue, their experiences and social histories.

A commitment to attend to learners' understandings offers an important critique of traditional forms of professional development. While pre-service teacher education courses often involve reflection on personal experiences or existing ideas about schooling, typical professional development workshops do not. Additionally, as situative learning theorists (Greeno, 1998; Lave & Wenger, 1991), our perspective on teacher learning encompasses affective and social, as well as cognitive, processes of transforming instructional practice.

Undoubtedly, teacher learning is a complex phenomenon. We are certainly not the first researchers to call for better theories of teacher learning. Throughout our study, we read the work of numerous scholars – primarily teacher educators – who have done important work on this topic and who contributed to our thinking. Nonetheless, the most well-articulated theories focus on novice teachers' development (e.g., Ammon & Hutcheson, 1989; Edwards, 2010; Korthagen, 2010; van Huizen, van Oers & Wubbels, 2005). These pieces offer insights, constructs, and images of the core phenomenon, but they did not fully illuminate what we saw in our empirical studies of experienced teachers' learning.

To guide our project, we turn to Jean Lave (1996)'s insight that adequate learning theories stipulate three things:

1. The basic *subject-world relationship*, accounting for how people interact with the world and come to understand it.
2. A *telos* that details what changes as people learn.
3. A description of *learning mechanisms* that describe how changes implied by learning come about (p. 156, attributed to Lave's conversations with Martin Packer).

Using Lave's "theory of learning theories" as a framework, our research team reviewed research on sociocultural theories of teacher learning. We found that studies varied in how explicitly they articulated these components. For example, regarding subject-world relationships, we found many studies focused on teacher cognition and behavior (e.g., Davis & Krajcik, 2005; Zwart et al., 2008), even when staking claims in the sociocultural terrain by highlighting the role of context. Some studies emphasized instructional practices' interactive dimensions, which inevitably highlight the socially negotiated nature of teaching (e.g., Connell, 2010; Guskey, 1986; Heyd-Metzuyanim et al., 2019). Many articulated the importance of teachers' subjectivities (e.g., Camburn, 2010) or took constructionist views, noting that teachers make their own sense of instructional events and practices (e.g., Bargagliotti & Anderson, 2017; Borko, 2004). Others addressed issues of structure and agency, with many emphasizing the role of teacher agency in their learning (e.g., Clarke & Hollingsworth, 2002; Waitoller & Artiles, 2016), underscoring that they make decisions about their actions. Still other studies highlighted cultural and institutional resources that shape practice and how teachers use those resources to make sense of instruction (e.g., Cameron et al., 2013; Clarke & Hollingsworth, 2002; Horn, 2005; Jurasaite-Harbison & Rex, 2010). Taken together, prior work shows teachers' relation to their knowledge domain as contextual, interactive, and interpretive, with their agency constrained by cultural and institutional resources.

Regarding how teachers' understandings change – the *telos* in Lave's framework – the shifts were often additive, with teacher learning occurring when they extended (as opposed to reorganized) their knowledge base (e.g., Bargagliotti & Anderson, 2017; Davis & Krajcik, 2005). Other studies moved closer to a conceptual change perspective, describing how particular beliefs (Meirink et al., 2009) or practices (Heyd-Metzyanim et al., 2019) evolved toward researchers' desired outcomes. When studies did not emphasize desired outcomes, the description of changes became understandably more ambiguous: Teachers developed "new forms of participation" (Hoffman-Kipp, 2003) or "more contextually responsive practices" (Attard, 2007); teachers learning together developed "changes in collective praxis" (Jahreie & Otteson, 2010). These constructivist accounts describe what is new when teachers learn but stop short of fleshing out the details of the nature of the change.

When analysts press closer into these details, they present teacher learning as a process of knowledge integration, what Anne Edwards (2010) described as new interpretations of the social world which relate to new actions. Likewise, Peter Dudley's (2013) research on teachers' learning in lesson study highlighted their development of new, classroom-specific knowledge about supporting student learning based on insights from the lesson study process. Importantly, the knowledge-integration view of telos implicitly critiques the additive view of learning; Ian Hardy's (2010) study of teacher learning noted a shift from an *attitude of compliance* in relation to desired practices to a *"researchly" reflective practice* focused on student learning. Compliance echoes an additive view of telos, while reflective practice points to knowledge integration.

Cognitive theories of learning often invoke mechanisms, such as the reorganization of schemas or beliefs, to describe how learning happens. But in most sociocultural studies we reviewed, mechanism was either omitted completely or was conflated with learning activities (e.g., specific details of interventions). For instance, teacher experimentation led to changes in practice (Meirink et al., 2009), or socialization into the norms of the intervention led to new ways of thinking about instruction (Boyd & Bloxham, 2014; van Es, 2011). Thematically, many studies emphasized reflection for supporting teacher change (Attard, 2007; Bargagliotti & Anderson, 2017; Meirink et al., 2009; Zwart et al., 2008), but how reflection related to teacher learning remained unspecified.

In the first section of this book, we describe teacher learning as a situative project of conceptual change. We organize our argument around Lave's theory of learning theories. In Chapter 1, we draw on research from the sociology of teaching to describe themes about subject-world relations that shape knowledge in, of, and for teaching practice (Cochran-Smith & Lytle, 1999). Although we aim for our theory to apply across schooling contexts, a dilemma of the situative perspective is that our work inherently points to the role of particular situations. For that reason, this chapter largely reflects teachers' current realities in U.S. schools; we keep this in mind to aid scholars in other places and times to use these ideas. In Chapter 2, the *telos* chapter, we describe teachers' learning of ambitious and equitable mathematics instruction as a conceptual change project, highlighting the nature of that change as teachers become more adept. In Chapter 3, we describe a key mechanism for conceptual change, moving beyond the truism of reflective practice to account for how reflection supports teachers' concept development.

References

Ammon, P., & Hutcheson, B. P. (1989). Promoting the development of teachers' pedagogical conceptions. *The Genetic Epistemologist, 17*(4), 23–29.

Attard, K. (2007). Habitual practice vs. the struggle for change: Can informal teacher learning promote ongoing change to professional practice? *International Studies in Sociology of Education, 17*(1–2), 147–162.

Bargagliotti, A. E., & Anderson, C. R. (2017). Using learning trajectories for teacher learning to structure professional development. *Mathematical Thinking and Learning, 19*(4), 237–259.

Borko, H. (2004). Professional development and teacher learning: Mapping the terrain. *Educational Researcher, 33*(8), 3–15.

Boyd, P., & Bloxham, S. (2014). A situative metaphor for teacher learning: The case of university tutors learning to grade student coursework. *British Educational Research Journal, 40*(2), 337–352.

Brousseau, G. (2006). *Theory of didactical situations in mathematics: Didactique des mathématiques, 1970–1990* (Vol. 19). Springer Science & Business Media.

Camburn, E. M. (2010). Embedded teacher learning opportunities as a site for reflective practice: An exploratory study. *American Journal of Education, 116*(4), 463–489.

Cameron, S., Mulholland, J., & Branson, C. (2013). Professional learning in the lives of teachers: Towards a new framework for conceptualising teacher learning. *Asia-Pacific Journal of Teacher Education, 41*(4), 377–397.

Clarke, D., & Hollingsworth, H. (2002). Elaborating a model of teacher professional growth. *Teaching and Teacher Education, 18*(8), 947–967.

Cochran-Smith, M., & Lytle, S. L. (1999). Chapter 8: Relationships of knowledge and practice: Teacher learning in communities. *Review of Research in Education, 24*(1), 249–305.

Connell, M. T. (2010). Framing teacher education: Participation frameworks as resources for teacher learning. *Pedagogies: An International Journal, 5*(2), 87–106.

Davis, E. A., & Krajcik, J. S. (2005). Designing educative curriculum materials to promote teacher learning. *Educational Researcher, 34*(3), 3–14.

diSessa, A. (2006). History of conceptual change research: Threads and fault lines. In Sawyer, R. K. (Ed.) (2005). *The Cambridge handbook of the learning sciences (pp. 265–281).* Cambridge University Press.

Dudley, P. (2013). Teacher learning in lesson study: What interaction-level discourse analysis revealed about how teachers utilised imagination, tacit knowledge of teaching and fresh evidence of pupils learning, to develop practice knowledge and so enhance their pupils' learning. *Teaching and Teacher Education, 34,* 107–121.

Edwards, A. (2010). How can Vygotsky and his legacy help us to understand and develop teacher education? In V. Ellis, A. Edwards, & P. Smagorinsky (Eds.), *Cultural-historical perspectives on teacher education and development.* Routledge.

Freudenthal, H. (2012). *Mathematics as an educational task.* Springer Science & Business Media.

Greeno, J. G. (1998). The situativity of knowing, learning, and research. *American Psychologist, 53*(1), 5.

Gresalfi, M., & Hand, V. M. (2019). Coordinating situated identities in mathematics classrooms with sociohistorical narratives: A consideration for design. *ZDM, 51*(3), 493–504.

Guskey, T. R. (1986). Staff development and the process of teacher change. *Educational Researcher, 15*(5), 5–12.

Gutstein, E. (2006). *Reading and writing the world with mathematics: Toward a pedagogy for social justice.* Taylor & Francis.

Hardy, I. (2010). Critiquing teacher professional development: Teacher learning within the field of teachers' work. *Critical Studies in Education, 51*(1), 71–84.

Heyd-Metzuyanim, E., Smith, M., Bill, V., & Resnick, L. B. (2019). From ritual to explorative participation in discourse-rich instructional practices: A case study of teacher learning through professional development. *Educational Studies in Mathematics, 101*(2), 273–289.

Hoffman-Kipp, P. (2003). Model activity systems: Dialogic teacher learning for social justice teaching. *Teacher Education Quarterly, 30*(2), 27–39.

Jahreie, C. F., & Ottesen, E. (2010). Learning to become a teacher. In V. Ellis, A. Edwards, and P. Smagorinsky (Eds.), *Cultural-historical perspectives on teacher education and development,* (pp. 131–145). London: Routledge.

Jurasaite-Harbison, E., & Rex, L. A. (2010). School cultures as contexts for informal teacher learning. *Teaching and Teacher Education, 26*(2), 267–277.

Korthagen, F. A. (2010). Situated learning theory and the pedagogy of teacher education: Towards an integrative view of teacher behavior and teacher learning. *Teaching and Teacher Education, 26*(1), 98–106.

Lampert, M. (2001). *Teaching problems and the problems of teaching.* Yale University Press.

Lave, J. (1996). Teaching, as learning, in practice. Mind, Culture, and Activity, 3(3), 149–164.

Lave, J., & Wenger, E. (1991). *Situated learning: Legitimate peripheral participation.* Cambridge University Press.

Meirink, J. A., Meijer, P. C., Verloop, N., & Bergen, T. C. (2009). Understanding teacher learning in secondary education: The relations of teacher activities to changed beliefs about teaching and learning. *Teaching and Teacher Education, 25*(1), 89–100.

Sinclair, N. (2004). The roles of the aesthetic in mathematical inquiry. *Mathematical Thinking and Learning, 6*(3), 261–284.

Van Es, E. A. (2011). A framework for learning to notice student thinking. In M. Sherin, V. Jacobs & R. Phillip (Eds.), *Mathematics Teacher Noticing: Seeing Through Teachers' Eyes* (pp. 134–151). Routledge

van Huizen, P., van Oers, B., & Wubbels, T. (2005). A Vygotskian perspective on teacher education. *Journal of Curriculum Studies, 37*(3), 267–290.

Waitoller, F. R., & Artiles, A. J. (2016). Teacher learning as curating: Becoming inclusive educators in school/university partnerships. *Teaching and Teacher Education, 59*, 360–371.

Zwart, R. C., Wubbels, T., Bolhuis, S., & Bergen, T. C. (2008). Teacher learning through reciprocal peer coaching: An analysis of activity sequences. *Teaching and Teacher Education, 24*(4), 982–1002.

1

THE CONDITIONS FOR TEACHER LEARNING AND NATURE OF TEACHER KNOWLEDGE

To understand teacher learning as situative concept development, we must first illuminate the subject-world relations between teachers-as-learners and teaching-as-knowledge. Subject-world relations are part of any learning theory, although they are seldom made explicit. For instance, behaviorism posits a sensing subject who encounters stimuli. The sensing subject's behaviors are shaped by the reinforcements or punishments of those stimuli. Alternatively, the cognitive revolution introduced a sensemaking subject who modeled the world through schemas – explanations of how things work that were revised through subsequent experiences. Our situative perspective expands a sensemaking perspective by blurring the boundary between subject and world, imagining concepts as extending beyond individuals into meanings in the world and related repertoires of cultural practice.

To describe how concepts about teaching do not merely reside in teachers' heads but are part of their worlds, we organize this chapter as follows: We illustrate our conceptual change perspective on teacher learning by discussing key findings in mathematics education research about different conceptions of teaching and learning. Then, we synthesize research on teacher learning and the sociology of teachers' work to argue that teachers' relationship to what they learn is socially embedded, ambiguous, and contested. We summarize these as *premises and corollaries* about teachers' subject-world relations.

Teacher Learning as Conceptual Change

To make ambitious and equitable mathematics instruction come alive in classrooms, teachers require deep, adaptable understandings of its aims. When teachers want to "teach mathematics for understanding," what does

DOI: 10.4324/9781003182214-2

understanding look like? How does it differ from the procedural knowledge most often emphasized in school mathematics? When teachers want to "engage all learners in meaningful content," what does *engagement* look like? What does it mean to identify the *strengths and needs* of all their students? What makes content *meaningful*? How is this different from typical instruction? Clearly, these questions about ambitious and equitable mathematics instruction require a tight coupling between teachers' understandings and repertoires of practice. At the same time, teachers' understandings are embedded in webs of meaning that are cultural, social, and historical. Given these entanglements, how do teachers make sense of these ideas in ways that reflect the goals of ambitious and equitable mathematics teaching?

Ambitious and equitable mathematics teaching requires substantial teacher learning on multiple fronts. Considering how mathematics has been taught historically – and how schools are typically organized – this learning requires that teachers transform taken-for-granted understandings of instruction. For this reason, we view learning to teach ambitiously as a *conceptual change project* that goes against the dominant culture of schooling (Cobb & Jackson, 2012; Stein & Wang, 1988). Because of its institutional incongruity, learning to teach ambitiously often becomes, to varying degrees, a cultural change project, too (Cook & Yanow, 1993; Horn & Gresalfi, 2021).

Emphasizing teachers' conceptual change moves away from common-sense ideas about instructional change. More commonly, instructional change is described as a problem of implementation. Schools adopt a new curriculum, teachers get training in new approaches, and then they "implement" them. But implementation is not a plug-and-play proposition; decades of research show that teachers' ideas about students, teaching, and mathematics – and local cultures of schooling – influence the nature of implementation (Coburn, 2001; Schoenfeld, 1988). These ideological and cultural forces shape how any innovation meets – or does not meet – its intended goals.

To illustrate this, consider a well-known task on proportional reasoning about orange juice concentrate (Cramer et al., 1993; Noelting, 1980; see Figure 1.1). Proportional reasoning is a foundational idea in mathematics, taken up in geometry (scaling polygons), algebra (slope), trigonometry (trig functions), and statistics (population sampling). The task embeds the mathematics in a potentially vivid context of tasting mixes of various strengths, with the aim of linking proportions to the real world. Students are given various ratios of orange juice concentrate and water, then prompted to explain which mixture would be the most "orangey."

While implementation necessarily involves adaptation, some adaptations undermine the intent of a tool, while others support (or even enhance) it. With the orangey task, we can imagine one teacher, Mr. Azure, adapting this task by directing students to compare the ratios, thus lowering the cognitive demand by proceduralizing the problem (Henningsen & Stein, 1997). This

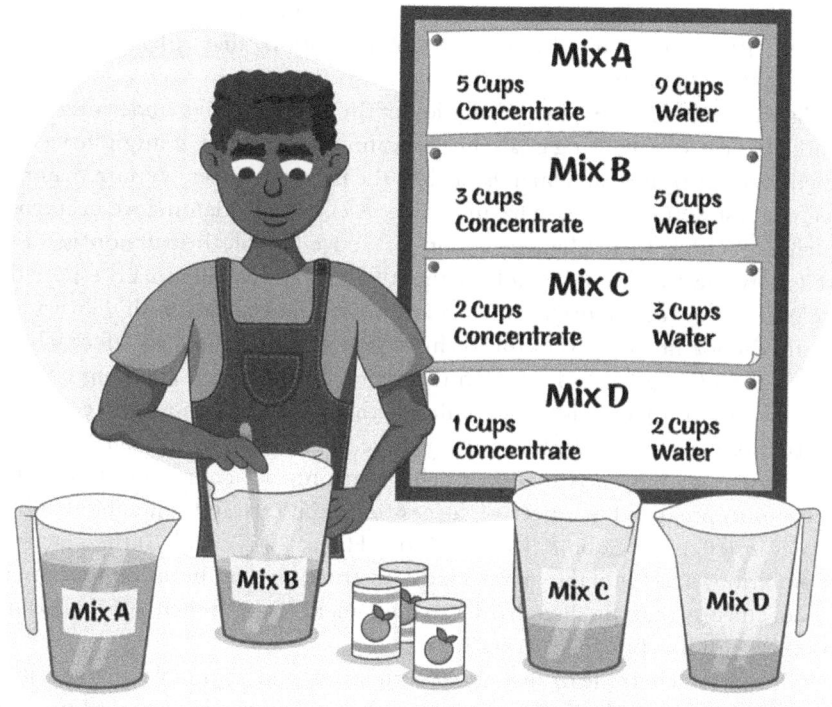

FIGURE 1.1 The Orangey Task.

adaptation simplifies the task and makes it quicker for students to complete; this may help Mr. Azure to keep up with his colleagues. Yet this hastening robs students of sensemaking opportunities. In contrast, Ms. Brown might adapt the task by changing "orange juice concentrate" (something not all students are familiar with) to lemonade powder from the school cafeteria. She might launch the problem by showing them powder and water, eliciting ideas about what will make the lemonade more "lemony," then invite students to work in small groups to decide which mixture is lemoniest, backing it with mathematical reasoning. Ms. Brown's adaptation responds to students' background knowledge while maintaining their role as mathematical sensemakers (Jackson et al., 2013). In this way, the nature of teachers' adaptations reveals their understandings of ambitious and equitable mathematics instruction – a distinction that James Spillane (2000) describes as "form versus function." In other words, using the task in and of itself – the *form* – does not ensure the realization of ambitious and equitable mathematics instruction – the *function*. This is a subtle point, particularly given simplistic notions of "fidelity of implementation" that position teachers as interchangeable actors who use curricula or instructional practices exactly as the designers intended. While there are certainly examples of adaptations that undermine instructional goals (e.g., Cohen, 1990), there are also examples of adaptations that enhance students' learning (e.g., Jackson et al., 2013).

As we hinted, Mr. Azure's adaptation of the task may have undermined the mathematical sensemaking goals, but we cannot ignore how it might have fulfilled other institutional demands, such as the need to "cover" certain material before a test or to adhere to a pacing guide. A common dilemma we hear from teachers seeking to implement ambitious and equitable instruction is, "Do we follow the curriculum or follow the students?" This dilemma is especially fraught for secondary teachers, who are often preoccupied with "covering" required content in preparation for high-stakes testing that can affect school funding, their reputations, or even their jobs. In this way, schooling culture and institutional demands often conflict with tenets of ambitious and equitable instruction. For these reasons, among those who study instructional change, the consensus opinion is that (1) teacher learning is a core – and frequently underconceptualized – aspect of successful ambitious and equitable mathematics instruction (Cobb & Jackson, 2012; Horn, 2005); and (2) the practices of ambitious and equitable mathematics instruction may be in opposition to school culture, making them harder to implement (Brantlinger, 2013; Chazan, 2000) or sustain (Louie & Nasir, 2014).

Clearly, teachers' learning about ambitious and equitable mathematics instruction is a complex project of conceptual and cultural change. Moreover, the indefinite, personal, practical knowledge of teaching (Clandinin & Connelly, 1996; Kumashiro, 2004) differs from studies of learning in well-structured domains like playing chess (e.g., Chase & Simon, 1973) or scientific reasoning (e.g., Hmelo-Silver et al., 2002). We draw this out to illustrate the complexity of subject-world relations for teacher learning. As our premises and corollaries will illustrate, teachers' relationships to what they learn are socially embedded, ambiguous, and contested.

Research on teacher learning does not always attend to the complex subject-world relations between teachers-as-learners and teaching-as-knowledge. Instead, teacher learning is typically studied as an outcome of educational experiences like professional development, pre-service education, or researcher-designed interventions. This acquire-apply perspective (Zeichner, 2010) has resulted in thin theory around notions of teacher learning – the details of the learning processes often remain unstated in studies, with an implicit theory that the educational experience causes changes in knowledge and beliefs that, in turn, change instruction (Biesta, 2015; Clarke & Hollingsworth, 2002). Although nuanced accounts of individual teacher sensemaking exist (Heaton, 2000; Lampert, 2001), the resulting frameworks have not made substantial inroads in the professional development literature, which tends to portray teacher-as-subject as relatively passive and teaching-as-knowledge as relatively certain, complete, and determinate.

Conceptual Change

To respond to the limitations of prior theory on teacher learning, we start with a conceptual change perspective. Conceptual change moves beyond simplistic forms of learning, like skill acquisition or additive changes to practice, and instead emphasizes the transformation of fundamental understandings (diSessa, 2005). This perspective is especially important for understanding teachers' learning in professional development, since working with practicing teachers means offering not *new* ideas about instruction, but ones that are *different* from teachers' existing ideas (Kennedy, 2016). While conceptual change research has not converged on one overarching theory, it does offer some important insights. First, as a theory of learning, conceptual change implicitly critiques notions of learning as acquisition, instead emphasizing that learners build new ideas in the context of old ones. Consequently, effective learning designs must surface and attend to those existing notions. Relatedly, conceptual change research emphasizes that learners are not blank slates; rather, they have various descriptions and explanations of the phenomena they encounter. In general, we find that professional development typically ignores teachers' extant understandings. This is a disconnect we seek to redress.

A conceptual change perspective on teacher learning asserts a fundamentally different subject–world relationship. Specifically, it emphasizes teachers' conceptual agency over their instruction. But ours is a situative theory of conceptual change; our analyses look beyond individual development to consider individuals-in-contexts. Accordingly, our conceptual change perspective emphasizes contextual resources and constraints for teachers' learning.

Some Premises About Teacher Learning

To capture the unique, characteristic relations between teacher-as-learner and teaching-as-knowledge that distinguish teacher learning from other kinds of learning, we synthesize themes from prior research as core premises about teacher learning, with an emphasis on the duality of teachers' conceptual agency and the contextual embeddedness of teachers' knowledge. As we suggested earlier, this consolidates themes from vast literatures on teacher learning and the sociology of teaching, which each offer insight into the situative nature of teacher knowledge, while making a playful nod to our own mathematical backgrounds.

The Action Premise: In Teaching, Action has Greater Currency Than Understanding

Teaching involves leading others through experiences designed to help them learn. These experiences are shaped by pedagogical actions, which are the most visible aspect of teaching. Consider a thought experiment: Imagine Ms. Crimson, who believes in the importance of spontaneity. Although she regularly arrives without clear plans of how to spend class time, she often reflects on classroom discussions and her students' ideas and understandings. Contrast this to Mx. Dandelion, who always arrives with detailed lesson plans, but does not reflect on their interactions with students. Which teacher would be viewed as "not doing their job"? Ms. Crimson's students, colleagues, and administrators might view her as negligent for not having prepared lessons; indeed, many U.S. schools require teachers to submit weekly lesson plans. Teachers who sit and think – even if they arrive at deeper understandings about student learning – do so even though it is not culturally or institutionally validated as a core task.

This thought experiment reveals two things. First, teachers' actions have greater urgency than reflection or understanding. Scholars who study teacher cognition stress the importance of this time constraint: Michael Eraut (1995) describes the time available for teachers' actions as limiting the possible modes of cognition; Frank Jansma and colleagues (1996) concur that the density of teachers' choices during lessons permit little time for reflection and often require immediate actions; Mary Kennedy (2006) describes the integrated habits that teachers use to respond to situations as they arise. The Action Premise highlights the primacy of embodied and actionable teacher knowledge in our culturally organized (and therefore taken-for-granted) understanding of teaching and what it means to know how to teach – in theoretical language, it shapes a hegemonic onto-epistemology of teaching. It also emphasizes the institutional valuation of action over reflection: That is, Mx. Dandelion's habits better align with typical institutional requirements than Ms. Crimson's, making action-over-reflection an easier path for teacher development.

The Alignment Corollary to the Action Premise

Thus, we arrive at a corollary to the Action Premise – the Alignment Corollary: Teachers are more likely to develop teaching actions that align with institutional arrangements. Working "against the grain" (Cochran-Smith, 1991) of schools may reflect teachers' ethical commitments (Chen et al., 2021), but the effort of doing more than schools are designed for can cause demoralization or burnout (Bartlett, 2004; Santoro, 2011). Even when misalignments are buffered through communities of like-minded colleagues, it is difficult for teachers to sustain counter-normative forms of instruction in the long-run (Huberman, 1993).

The Alignment Corollary can be thought of as institutions providing well-worn paths for instructional practice that are easier to follow, with clear toeholds that demand less strategy and effort. Indeed, Étienne Wenger (1998) notes that alignment is a hallmark of belonging to a community of practice, which is a source of identity. For teachers, how they align with their institutions also shapes who they are becoming. As people engage with practices in their social world, they question or accept the norms they encounter. Barbara Jaworski (2006) speaks of *critical alignment*, wherein teachers align with aspects of practice while critically questioning their own roles and purposes within that participation. Teachers who engage other instructional approaches find themselves forging new paths, adding the labor of trailblazing to the already considerable work of teaching.

The Interpretation Premise: Teachers' Actions are Shaped by Their Interpretations of Instructional Moments

Classrooms are complex social spaces. Things often occur outside of teachers' awareness. A student may feel upset because they had an argument with someone before class; they may have a crush on a classmate and feel distracted or self-conscious; they may have forgotten their homework and feel unprepared; or they may have conspired with other classmates to bully someone without drawing the teacher's notice. Yet, despite teachers' incomplete knowledge of all that happens during their lessons, as the Action Premise suggests, they must continue to *do* things. Teachers are thus processing information about the classroom constantly – often with only somewhat deliberate thought, using what psychologist Seymour Epstein (1990) called "the experiential body-mind system." In other words, their actions and reactions are often based on unconscious images, feelings, values, and needs (Kennedy, 2006; Lampert, 2001). We have heard teachers describe what informs their instructional actions as closest to the kind of instinct involved in balancing on a bicycle. Often, they intuit the need to adjust instruction without explicit thought. At other times, an unexpected jolt may require more intentional action, in which case teachers' responses may be based on quick deliberation. To whatever extent they are instinctual or explicit, their actions represent their interpretations of classroom activity.

To illustrate this, let's take two examples of ambiguous moments that require teacher interpretation. First, Ms. Emerald discusses fraction addition with her students. They are trying to make sense of the sum of ⅛ and ¼. After drawing some diagrams together, a student raises her hand and says, "Three! There would be 3." Ms. Emerald has a choice, depending on her interpretation of that answer. On one hand, the answer is incorrect, since ⅛ + ¼ is not 3. Alternatively, she can interpret "three" as the beginning of "⅜" and invite the student to elaborate on *what* three she sees.[1] Second, Mr. Fuchsia notices that a typically enthusiastic student is rather subdued one day. Again, he has many options, depending on his interpretation of the student's low energy. Mr. Fuchsia could deliberately call on the student to try to re-engage her in the lesson; he could find a private moment to speak to her and ask if she is okay; or he could choose to ignore this anomalous day, waiting to see how things go the next day.

In both cases, the teachers' actions are shaped by a combination of their instinctual and deliberate interpretations of the situation. If Ms. Emerald's instruction is organized around correct answers instead of student thinking, she might interpret "3" as incorrect and move on. If Mr. Fuchsia has a goal of student participation in every class, he might be more inclined to call on the subdued student.

At the same time, multiple social and historical contexts also shape their interpretations. Teachers' lived experiences and sociocultural identities influence their interpretations of interactional moments (Chen, 2020; Self & Stengel, 2020). For instance, Ms. Emerald and her student may be from a historically marginalized group; this shared identity may inform her interpretation of the student as competent and capable of elaboration (Battey et al., 2018). Mr. Fuchsia may have suffered from depression as a teenager; he may wonder if his student is experiencing something similar, leading him to check in privately.

On an institutional level, situational resources may tip teachers' interpretations and actions in one direction over another: Ms. Emerald may be in a teacher community invested in supporting student thinking (Horn, 2010), giving her models of pursuing her student's idea and building on her sensemaking. Mr. Fuchsia may be hosting a student teacher, leading him to demonstrate what he believes is the most appropriate response; his school may have an embedded health clinic, giving him resources to refer the student for a depression screening.

The Utility Premise: Teachers Revise Understandings When Current Understandings no Longer Seem Useful

Teachers' press for action over understanding contributes to a pragmatic view of instructional change. Time and again, educational research from different traditions shows that practicality governs teachers' activity (Doyle & Ponder, 1977; Janssen et al., 2015). For instance, in work examining mathematics teachers' justifications for their actions, researchers described teachers' reasoning as shaped by an underlying logic of "practical rationality" (Herbst & Chazan, 2011). On a larger timescale, educational historian Jack Schneider (2014) found that research ideas that successfully became a part of teaching practice share several characteristics: They are visible to teachers, friendly to their worldview, easily shared, and practical within the structure of K-12 schooling. This last quality suggests the importance of practicality in teacher learning. Similarly, in a longitudinal ethnographic study of pre-service teachers' learning, teachers applied a *utility filter* to ideas offered in professional education, investing in practices that seemed useful and rejecting those that did not (Nolen et al., 2014).

As the notion of *utility filter* suggests, a practical view of instruction is typically concerned about what works. When this emphasis on pragmatism couples with what sociologist Dan Lortie (1975/2020) called the *presentism* and *conservatism* that characterize teachers' orientation to time, utility leads them to take up instructional practices that satisfy immediate, short-term goals, such as Mr. Azure's choice to prioritize his instructional pace over students' sense-making opportunities. Indeed, many studies show teachers explicitly rejecting new practices because they stretch the limit of existing classroom management techniques and increase the complexity of teaching (Cohen, 2011; Doyle, 1988). One way of interpreting this tendency is that *the most frequent and salient feedback teachers get about whether their instruction is "working" is the pragmatic question of whether their lessons are running smoothly.*

The Smoothness Corollary to the Utility Premise

The salience of smoothness yields a corollary to the Utility Premise: Namely, that *smoothness, as feedback for teacher learning, offers limited information for revising teachers' understandings and actions.* The Smoothness Corollary is a byproduct of the typical classroom arrangement, with one teacher vastly outnumbered by their students, whose cooperation is required for instruction to proceed (Cohen, 2011).

"Smooth-running lessons" happen in multiple ways; not all are equally good for students' mathematical learning. Undoubtedly, there is an institutional imperative for smoothness, and administrator feedback tends to focus on classroom management and organization (Rigby et al., 2017). In interviews, administrators often describe good teachers in terms of classroom management, whether it stems from their charismatic personalities or competence; they seldom express interest in the against-the-grain, transformative teaching that ambitious and equitable mathematics teaching invites (Pinto et al., 2012).

But smooth lessons do not always indicate deep student learning; in fact, meaningful classroom learning often involves debate, deliberation, uncertainty, and emotion (Cooper, 2014). There are many reasons lessons might be smooth but not support deep learning. For instance, teachers are often encouraged to organize elaborate systems for student compliance, but with rote mathematics content. In contrast, lessons can be less "smooth," yet mathematically rich and highly engaging, as Jo Boaler (2002) documented in her comparison of Amber Hill and Phoenix Park high schools. Amber Hill's teacher-centered instruction was fast-paced, suggesting smooth lessons, but teachers emphasized rote mathematics. Phoenix Park's less structured, student-centered instruction emphasized problem solving, which was often bumpier – indeed, Boaler observed various ways students appeared "off-task" during classes, including side conversations and periods of inactivity. Nonetheless, the students from each school performed comparably on the United Kingdom's

GSCE high school mathematics exam – even though only Amber Hill's "top set" students took it, while Phoenix Park drew from "mixed ability" levels. This suggests that the orderly Amber Hill lessons did not yield better learning outcomes than the lively, unpredictable lessons at Phoenix Park. In fact, the opposite seemed to be true: Phoenix Park's approach resulted in more learning for a wider swath of students. Moreover, students' affinity for mathematics was greater at Phoenix Park, as they demonstrated more ownership of their mathematical knowledge and more flexible understandings of key concepts (Boaler & Selling, 2017).

If smooth lessons do not necessarily indicate optimal learning environments, then the feedback that administrators value orients teachers to the wrong thing, which creates misalignment with the goals of schooling and those of ambitious and equitable instruction. We cannot overstate the importance of these institutional arrangements: The feedback people orient to defines what counts as success and where they put effort to appear competent. Just as the book *Moneyball* (Lewis, 2003) documented how baseball statistics like stolen bases and batting averages are poor indicators for successful players (and how overlooked statistics, like on-base percentage and slugging percentage, have greater predictive power), the profession of teaching needs to re-orient itself to the right lodestar to move forward. To use a metaphor from mathematical modeling (Silver, 2012), mixing up the signal of learning with the noise of smoothness leaves teachers paying attention to the wrong feedback, which then impedes teachers' learning of ambitious and equitable mathematics instruction. Since this form of teaching increases instructional complexity – very likely adding "bumps" to lessons – it adds an extra challenge to teachers' learning as they must let go of commitments to certain forms of smoothness.

If smoothness is the noise and student learning is the signal, how do we help teachers orient to better feedback as they gauge the utility of instructional practices? Given that teachers' primary feedback is from students (Metz, 1993), such a shift would require new feedback systems. Generally, feedback is crucial to helping people learn (Bransford et al., 1999), since it offers insight into what is working or not working about what they are doing. In the most basic, behaviorist sense, feedback reinforces actions. When administrators praise teachers for having "well-behaved classes" and make no comment on the quality of the content, this feedback reinforces the idea of student compliance as a valued aspect of lessons. When students balk at collaborating in groups, this feedback might interrupt the smoothness of a lesson, leading teachers to conclude that convincing students to work together productively is not worth the effort.

If *smoothness of a lesson* is problematic feedback for teachers because it is not adequately tethered to information about student learning, then what are alternatives? As people who study teachers' workplace learning, we observe that U.S. teachers have few places for meaningful feedback, especially after their first few years on the job. Indeed, many of the experienced teachers in our study told us we were the first to observe their classrooms in years. This is common in the U.S., where experienced teachers are typically observed by administrators every few years; if the teacher has a good reputation, these observations can become perfunctory. Even so, administrator observations are usually evaluative, inviting safe lessons with smooth performances – high-stakes observations are not amenable to critical feedback. Moreover, evaluation frameworks often are generic, with no connection to the goals of ambitious and equitable mathematics teaching (Rigby et al., 2017). Again, many teachers in our study shared that they had no observations (outside colleagues or the PDO leaders) that engaged the mathematics they were teaching. For these reasons, teacher observations do not typically provide feedback that supports teacher learning toward ambitious and equitable instruction.

In-class assessments offer another potential form of feedback that is more closely tied to student learning. Ideally, if students do not learn as expected, teachers can revisit their instruction and revise their practice. Indeed, strong formative assessment practices are connected to tremendous gains in student learning (Black & Wiliam, 2009). But, recalling our critique of simplistic notions of implementation, teachers' understandings shape the efficacy of formative assessment – another place where they exert conceptual agency in interpreting instruction. Unfortunately, students' unsuccessful performance on in-class assessments is often interpreted as a natural consequence of the unequal distribution of mathematical ability or differences in student motivation (Horn, 2007). In their study of instructional improvement in middle-school mathematics, Kara Jackson and colleagues (2017) found that most teachers did not view all students as capable of participating in rigorous mathematics. Instead, teachers attributed at least some of students' difficulty to inherent traits or deficits in their families or communities (see also Copur-Gencturk et al., 2020). As a result, teachers lowered the cognitive demand of activities if they perceived that students were facing difficulty, thus moving away from the aims of ambitious instruction (Wilhelm, 2014). While feedback from formative assessment can support teacher learning, Jackson and colleagues' work underscores the importance of supporting teachers' productive interpretations of that feedback.

So where does that leave meaningful feedback for teaching? Unless teachers have ways of connecting students' learning outcomes to their instruction while interrogating their assumptions about their students' capabilities, in-class assessments may only serve to reinforce narrow views of who can and cannot do mathematics. Additionally, high-stakes standardized tests that purport to measure student learning are often limited in what they make available for teachers to learn from (Garner, 2018) and are often neither timely nor specific enough to the curriculum to be useful (Kerr et al., 2006). Finally, teachers might receive feedback from colleagues during professional development or co-planning sessions. As we shall elaborate next in the Asynchronicity of

Reflection Premise, this feedback suffers from the limited view their peers might have into their classroom dynamics, character, or challenges, particularly if they teach in a different setting (Kennedy, 2010).

The Utility Premise and Smoothness Corollary illustrate teachers' conceptual agency in their instructional practice. If student learning is a primary goal of teaching, then what teachers consider "useful" depends on how they know that learning is happening. As we have argued, teachers' concepts of *student learning* are especially critical to utility judgments. Additionally, this premise and its corollary highlight how teachers' knowledge is embedded in the institutional context of teaching.

The Asynchronicity of Reflection Premise: The Process of Deliberately Revising Understandings and Planning New Actions Generally Happens Separately from Active Instruction

If teachers need feedback to spur changes in their instruction, this feedback requires reflection. While some reflection can happen during active instruction, some of the richer revisions happen asynchronously, either while planning lessons (pre-active reflection) or debriefing (post-active reflection). Indeed, a powerful aspect of Japanese Lesson Study (Lewis et al., 2006) is that it links the pre-active, active, and post-active lesson phases to help teachers understand instruction more deeply.

Typically, however, the asynchronous nature of reflection leaves teachers with limited fodder for sensemaking. Classrooms are socially dense spaces, filled with people, interactions, and relationships, so teachers' perception of what happens in any given lesson is necessarily partial (P. Jackson, 1968/2005). The substance for *post hoc* reflection – a sense of "smoothness," particular moments, remembered dialog – are both incomplete and filtered through teachers' interpretations. Although teacher noticing has been given substantial attention in the mathematics education literature (Sherin et al., 2011), few scholars have attended to the role teachers' subjectivities play in shaping the nature of that noticing.

Scholars who analyze the political nature of noticing add a critical dimension to this work. For instance, Adam Lefstein and Julia Snell (2011) argue that what teachers notice (and ignore) is political as well as cognitive, leading to biases in how teachers understand phenomena. The authors highlight the power relations that are implicated in what teachers notice, specifically that the social organization of teaching privileges certain kinds of seeing and devalues other kinds. Through a careful case study, Nicole Louie (2018) goes beyond the cognitive orientation of teacher noticing and illustrates its cultural and ideological dimensions, highlighting the contested, political nature of what teachers attend to and ignore (see also Baldinger, 2017). Grace Chen

(2020) extends this work, describing noticing as an affective as well as cognitive process that invokes teachers' historicized selves, making it effectively a "form of political labor" (Ahmed, 2017, p. 32, as cited in Chen, 2020).

This socially situated view of noticing reflects our situative view of teacher learning. The Interpretation Premise asserts that teachers are subjective actors who notice different details of lessons and filter them through their own experiences and sociohistorical identities – their gender, race, language, class, religion, and so on. What they recall is necessarily partial and already shaped by their own sense of teaching, learning, and mathematics. Ultimately, these observational limitations make reflection insufficient for supporting the transformative changes ambitious and equitable instruction requires.

Because teachers filter what they observe to reflect on later, the Asynchronicity of Reflection Premise shows the conceptual agency they exert in their own learning and sensemaking. Moreover, their reflections are shaped by their school contexts, with conditions like class size, lesson length, and other resources making certain issues more or less salient in the first place.

The Competing Visions of Quality Premise

While documents like the National Council of Teachers of Mathematics' *Principles and Standards* (2000) and *Principles to Actions* (2014) describe visions of ambitious and equitable mathematics instruction (which we outline in Chapter 2), there are competing visions of what constitutes good teaching. Ultimately, visions of good teaching are highly consequential for teachers' learning and motivation to change their practice. Charles Munter and Richard Correnti (2017) found teachers' instructional visions positively correlate with their growth in ambitious and equitable instruction, more so than their mathematical knowledge for teaching.

But as Kevin Kumashiro (2004) argues, "good teaching" is not a neutral concept. In fact, common sense notions of good teaching often conflict with goals of equity and liberation. In mathematics education, these competing visions are clearly visible. The U.S. Math Wars that raged in the 1990s and 2000s reflected competing ideas of what it means to know math, how it should be taught – and, as educational anthropologist Lisa Rosen (2000) argued – who has a right to learn it (Schoenfeld, 2004).

This points to the distinction between the organization of schooling – which implicates systems like tracking and questions about whether students with disabilities can learn in mainstream classrooms – and teachers' pedagogical choices. Obviously, as Mr. Azure illustrates, school organization shapes teachers' choices. Nonetheless, much research in teacher learning accounts for the latter more than the former, envisioning teachers as hyperagentic learners, even though institutional contexts certainly shape notions of good teaching, which teachers either work with or against. Gaining traction with counternormative forms of instruction requires a different kind of teacher learning, so teachers' contexts – especially the messages in their environments about what good teaching is (see the Alignment Corollary) – matter for their learning.

The Competing Visions of Quality Premise also points to teachers' conceptual agency in their learning, as well as the crucial role of context. If teachers are asked to implement instructional practices that counter their own sense of good teaching, they may try to undermine such efforts (Coburn, 2001) or even leave the profession (Santoro, 2011). Messages about desired practice are often institutionally embedded, sometimes through evaluation systems that pressure teachers to conform. The extent to which teachers' personal sense of good teaching aligns with messages in their contexts shapes what they learn and how they ultimately make sense of it.

Summary

In this chapter, we have argued that teacher learning has a unique character given the institutional nature of their work and the ambiguity of teaching knowledge. We have offered five premises and two corollaries, which we will refer to in subsequent chapters, as we illustrate how mathematics teachers learn to teach ambitiously. We summarize them in Table 1.1 below for easy reference. We claim that these premises and their corollaries encapsulate a duality in subject-world relations between teachers-as-learners and teaching-as-knowledge – namely, that teachers exert conceptual agency over their practice while, at the same time, teaching concepts are deeply embedded in

TABLE 1.1 A summary of the premises that shape subject-world relations between teachers and teaching knowledge, which, in turn, shape their instruction.

Action	In teaching, action has greater currency than understanding.
Alignment	Teachers are more likely to develop actions aligned with their institutions and contexts.
Interpretation	Teachers' actions are shaped by their interpretations of instructional moments.
Utility	Teachers revise understandings when current understandings no longer support productive action.
Smoothness	Smoothness of lessons, as feedback for teacher learning, offers limited information for revising teachers' understandings and actions.
Asynchronicity of Reflection	The process of deliberately revising understandings and planning new actions generally happens separately from active instruction.
Competing Visions of Quality	Visions of quality teaching, which vary in their commitments, are distributed across people and institutions. Teachers must navigate these sometimes contradictory messages as a part of their sensemaking.

(and thus bound by) broader contexts, particularly the institutional setting of schools, which shape teachers' knowing. In this way, teachers' relationship to what they learn is socially embedded, ambiguous, and contested.

Note

1 This example is similar to one described about Magdalene Lampert's teaching in Cazden, C. B. (1988). *Classroom Discourse: The Language of Teaching and Learning*. Heinemann.

References

Ahmed, S. (2017). *Living a feminist life*. Durham, NC: Duke University Press.

Baldinger E. M. (2017) "Maybe It's a Status Problem." Development of mathematics teacher noticing for equity. In E. Schack, M. Fisher & J. Wilhelm (Eds.), *Teacher noticing: Bridging and broadening perspectives, contexts, and frameworks. Research in mathematics education*. Springer, Cham. https://doi.org/10.1007/978-3-319-46753-5_14

Bartlett, L. (2004). Expanding teacher work roles: a resource for retention or a recipe for overwork?. Journal of Education Policy, 19(5), 565–582.

Battey, D., Leyva, L. A., Williams, I., Belizario, V. A., Greco, R., & Shah, R. (2018). Racial (mis) match in middle school mathematics classrooms: Relational interactions as a racialized mechanism. *Harvard Educational Review, 88*(4), 455–482.

Biesta, G. (2015). On the two cultures of educational research, and how we might move ahead: Reconsidering the ontology, axiology and praxeology of education. *European Educational Research Journal, 14*(1), 11–22.

Boaler, J. (2002). *Experiencing school mathematics: Traditional and reform approaches to teaching and their impact on student learning*. Routledge.

Boaler, J., & Selling, S. K. (2017). Psychological imprisonment or intellectual freedom? A longitudinal study of contrasting school mathematics approaches and their impact on adults' lives. *Journal for Research in Mathematics Education, 48*(1), 78–105.

Black, P., & Wiliam, D. (2009). Developing the theory of formative assessment. *Educational Assessment, Evaluation and Accountability (formerly: Journal of Personnel Evaluation in Education), 21*(1), 5.

Bransford, J. D., Brown, A. L., & Cocking, R. R. (1999). How people learn: Brain, mind, experience, and school. National Academy Press.

Brantlinger, A. (2013). Between politics and equations: Teaching critical mathematics in a remedial secondary classroom. *American Educational Research Journal, 50*(5), 1050–1080.

Cazden, C. B. (1988). *Classroom discourse: The language of teaching and learning*. Heinemann.

Chase, W. G., & Simon, H. A. 1973. Perception in chess. *Cognitive Psychology, 4*: 55–81.

Chazan, D. (2000). *Beyond formulas in mathematics and teaching: Dynamics of the high school algebra classroom*. Teachers College Press.

Chen, G. A. (2020). "That's obviously really insensitive:" Attuning to marginalization in a parent-teacher encounter. *Cognition and Instruction, 38*(2), 153–178.

Chen, G. A., Marshall, S. A., & Horn, I. S. (2021). 'How do I choose?': Mathematics teachers' sensemaking about pedagogical responsibility. Pedagogy, Culture & Society, 29(3), 379–396.

Clandinin, D. J., & Connelly, F. M. (1996). Teachers' professional knowledge landscapes: Teacher stories – stories of teachers – school stories – stories of schools. *Educational Researcher, 25*(3), 24–30.

Clarke, D., & Hollingsworth, H. (2002). Elaborating a model of teacher professional growth. *Teaching and Teacher Education, 18*(8), 947–967.

Cobb, P., & Jackson, K. (2012). Analyzing educational policies: A learning design perspective. *Journal of the Learning Sciences, 21*(4), 487–521.

Coburn, C. E. (2001). Collective sensemaking about reading: How teachers mediate reading policy in their professional communities. *Educational Evaluation and Policy Analysis, 23*(2), 145–170.

Cochran-Smith, M. (1991). Learning to teach against the grain. *Harvard Educational Review, 61*(3), 279–311.

Cohen, D. K. (1990). A revolution in one classroom: The case of Mrs. Oublier. *Educational Evaluation and Policy Analysis, 12*(3), 311–329.

Cohen, D. K. (2011). *Teaching and its predicaments.* Harvard University Press.

Cook, S. N., & Yanow, D. (1993). Culture and organizational learning. *Journal of Management Inquiry, 20*(4), 362–379.

Cooper, K. S. (2014). Eliciting engagement in the high school classroom: A mixed-methods examination of teaching practices. *American Educational Research Journal, 51*(2), 363–402.

Copur-Gencturk, Y., Cimpian, J. R., Lubienski, S. T., & Thacker, I. (2020). Teachers' bias against the mathematical ability of female, Black, and Hispanic students. *Educational Researcher, 49*(1), 30–43.

Cramer, K., Post, T., & Graeber, A. O. (1993). Connecting research to teaching: Proportional reasoning. *The Mathematics Teacher, 86*(5), 404–407.

diSessa, A. A. (2005). A History of Conceptual Change Research. In R.K. Sawyer (Ed.), *The Cambridge Handbook of the Learning Sciences* (pp. 265–282), Cambridge University Press.

Doyle, W., & Ponder, G. A. (1977). The practicality ethic in teacher decision-making. *Interchange, 8*(3), 1–12.

Epstein, S. (1990). Cognitive-experiential self-theory. In L. A. Pervin (Ed.), *Handbook of personality, theory and research* (pp. 165–192). The Guilford Press.

Eraut, M. (1995). Schon shock: A case for refraining reflection-in-action? *Teachers and Teaching, 1*(1), 9–22.

Garner, B. (2018). *Data use for instructional improvement: Tensions, concerns, and possibilities for supporting ambitious and equitable instruction* (Doctoral dissertation).

Heaton, R. M. (2000). *Teaching mathematics to the new standard: Relearning the dance* (Vol. 15). Teachers College Press.

Henningsen, M., & Stein, M. K. (1997). Mathematical tasks and student cognition: Classroom-based factors that support and inhibit high-level mathematical thinking and reasoning. *Journal for Research in Mathematics Education, 28*(5), 524–549.

Herbst, P., & Chazan, D. (2011). Research on practical rationality: Studying the justification of actions in mathematics teaching. *The Mathematics Enthusiast, 8*(3), 405–462.

Hmelo-Silver, C. E., Nagarajan, A., & Day, R. S. (2002). "It's harder than we thought it would be": A comparative case study of expert–novice experimentation strategies. *Science Education, 86*(2), 219–243.

Horn, I. S. (2005). Learning on the job: A situated account of teacher learning in high school mathematics departments. Cognition and Instruction, 23(2), 207–236.

Horn, I. S. (2007). Fast kids, slow kids, lazy kids: Framing the mismatch problem in mathematics teachers' conversations. *The Journal of the Learning Sciences, 16*(1), 37–79.

Horn, I. S. (2010). Teaching replays, teaching rehearsals, and re-visions of practice: Learning from colleagues in a mathematics teacher community. *Teachers College Record, 112*(1), 225–259.

Horn, I. S., & Gresalfi, M. (2021). Broadening participation in mathematical inquiry: A problem of instructional design. In R. Duncan & C.A. Chinn (Eds.), *International handbook of inquiry and learning* (pp. 311–324). Routledge.

Huberman, M. (1993). The model of the independent artisan in teachers' professional relations. *Teachers' Work: Individuals, Colleagues, and Contexts*, 11–50.

Jackson, K., Garrison, A., Wilson, J., Gibbons, L., & Shahan, E. (2013). Exploring relationships between setting up complex tasks and opportunities to learn in concluding whole-class discussions in middle-grades mathematics instruction. *Journal for Research in Mathematics Education*, *44*(4), 646–682.

Jackson, K., Gibbons, L., & Sharpe, C. J. (2017). Teachers' views of students' mathematical capabilities: Challenges and possibilities for ambitious reform. *Teachers College Record*, *119*(7), 1–43.

Jackson, P. (1968/2005). *Life in classrooms*. Routledge.

Janssen, F., Westbroek, H., & Doyle, W. (2015). Practicality studies: How to move from what works in principle to what works in practice. *Journal of the Learning Sciences*, *24*(1), 176–186.

Jansma, F., Wubbels, T., Korthagen, F., & Dolk, M. (1996). *The relation between teacher thought and behavior: Implications for teacher training*. Paper presented at the annual meeting of the American Educational Research Association, New York.

Jaworksi, B. (2006). Theory and practice in mathematics teaching development: Critical inquiry as a mode of learning in teaching. *Journal of Mathematics Teacher Education*, *9*, 187–211.

Kennedy, M. M. (2006). Knowledge and vision in teaching. *Journal of Teacher Education*, *57*(3), 205–211.

Kennedy, M. M. (2010). Attribution error and the quest for teacher quality. Educational Researcher, 39(8), 591–598.

Kennedy, M. M. (2016). How does professional development improve teaching? *Review of Educational Research*, *86*(4), 945–980.

Kerr, K. A., Marsh, J. A., Ikemoto, G. S., Darilek, H., & Barney, H. (2006). Strategies to promote data use for instructional improvement: Actions, outcomes, and lessons from three urban districts. *American Journal of Education*, *112*(4), 496–520.

Kumashiro, K. K. (2004). Uncertain beginnings: Learning to teach paradoxically. *Theory into Practice*, *43*(2), 111–115.

Lampert, M. (2001). *Teaching problems and the problems of teaching*. Yale University Press.

Lefstein, A., & Snell, J. (2011). Professional vision and the politics of teacher learning. *Teaching and Teacher Education*, *27*(3), 505–514.

Lewis, C., Perry, R., & Murata, A. (2006). How should research contribute to instructional improvement? The case of lesson study. *Educational Researcher*, *35*(3), 3–14.

Lewis, M. (2003). *Moneyball: The art of winning an unfair game*. W.W. Norton.

Lortie, D. C. (1976/2020). *Schoolteacher: A sociological study*. University of Chicago Press.

Louie, N. L. (2018). Culture and ideology in mathematics teacher noticing. *Educational Studies in Mathematics*, *97*(1), 55–69.

Louie, N., & Nasir, N. (2014). Derailed at Railside. In N. Nasir, C. Cabana, B. Shreve, E. Woodbury, & N. Louie (Eds.) *Mathematics for Equity: A Framework for Successful Practice*. pp. 187-206. Teachers College Press.

Metz, M. H. (1993). Teachers' ultimate dependence on their students. In J. W. Little, & M. W. McLaughlin (Eds.), *Teachers work: Individuals, colleagues, and contexts* (pp. 104–137). New York: Teachers College Press.

Munter, C., & Correnti, R. (2017). Examining relations between mathematics teachers' instructional vision and knowledge and change in practice. *American Journal of Education, 123*(2), 171–202.

National Council of Teachers of Mathematics. (2000). *Principles and standards for school mathematics*. National Council of Teachers of Mathematics.

National Council of Teachers of Mathematics. (2014). *Principles to actions: Ensuring mathematics success for all*. National Council of Teachers of Mathematics.

Noelting, G. (1980). The development of proportional reasoning and the ratio concept Part I – Differentiation of stages. *Educational Studies in Mathematics, 11*(2), 217–253.

Nolen, S. B., Ward, C. J., & Horn, I. S. (2014). Changing practice(s). *Teacher Motivation: Theory and Practice*, 167.

Pinto, L. E., Portelli, J. P., Rottmann, C., Pashby, K., Barrett, S. E., & Mujawamariya, D. (2012). Charismatic, competent or transformative? Ontario school administrators' perceptions of "good teachers." *Journal of Teaching and Learning, 8*(1). https://doi.org/10.22329/jtl.v8i1.3052.

Rigby, J. G., Larbi-Cherif, A., Rosenquist, B. A., Sharpe, C. J., Cobb, P., & Smith, T. (2017). Administrator observation and feedback: Does it lead toward improvement in inquiry-oriented math instruction? *Educational Administration Quarterly, 53*(3), 475–516.

Rosen, L. (2000) Calculating concerns: The politics or representation in California's "Math Wars." Unpublished doctoral dissertation. University of California, San Diego.

Santoro, D. A. (2011). Teaching's conscientious objectors: Principled leavers of high-poverty schools. *Teachers College Record, 113*(12), 2670–2704.

Schneider, J. (2014). *From the Ivory Tower to the schoolhouse: How scholarship becomes common knowledge in education*. Harvard Education Press.

Schoenfeld, A. H. (1988). When good teaching leads to bad results: The disasters of 'well-taught' mathematics courses. Educational Psychologist, 23(2), 145–166.

Schoenfeld, A. H. (2004). The math wars. *Educational Policy, 18*(1), 253–286.

Self, E. A., & Stengel, B. S. (2020). *Toward anti-oppressive teaching: Designing and using simulated encounters*. Harvard Education Press.

Silver, N. (2012). *The signal and the noise: Why so many predictions fail – but some don't*. Penguin.

Sherin, M., Jacobs, V., & Philipp, R. (Eds.). (2011). *Mathematics teacher noticing: Seeing through teachers' eyes*. Routledge.

Spillane, J. P. (2000). Cognition and policy implementation: District policymakers and the reform of mathematics education. *Cognition and Instruction, 18*(2), 141–179.

Stein, M. K., & Wang, M. C. (1988). Teacher development and school improvement: The process of teacher change. *Teaching and Teacher Education, 4*(2), 171–187.

Wenger, E. (1998). *Communities of practice: Learning, meaning, and identity*. Cambridge University Press.

Wilhelm, A. G. (2014). Mathematics teachers' enactment of cognitively demanding tasks: Investigating links to teachers' knowledge and conceptions. *Journal for Research in Mathematics Education, 45*(5), 636–674.

Zeichner, K. (2010). Rethinking the connections between campus courses and field experiences in college- and university-based teacher education. *Journal of Teacher Education, 61*(1–2), 89–99.

2

HOW CONCEPTS CHANGE AS TEACHERS LEARN AMBITIOUS AND EQUITABLE MATHEMATICS INSTRUCTION

With Grace A. Chen

To understand teacher learning as situated concept development, we need to consider the *telos* of teachers' learning – that is, what changes as they learn ambitious and equitable instruction. Prior research has documented that, to successfully manage this kind of teaching, teachers must shift their understandings away from traditional strategies toward new methods that increase ambiguity, uncertainty, and risk. But what else changes as they do so?

In this chapter, we begin by describing three key shifts for developing ambitious and equitable mathematics instruction. These address the organization of instructional activities; what math class sounds like; and who belongs in math class. For analytic clarity, we parcel these out as individual shifts; however, they are deeply interrelated. Then, in line with our situative perspective, we describe the *conceptual resources* that anchor common-sense understandings of instruction that may lie beyond individual teachers, pointing to the ways concepts span across people, activities, and contexts (Lave, 1988). Finally, we describe another important aspect of teachers' telos: the development of integrated concepts for teaching. We end the chapter by briefly discussing how these intertwine to account for what changes as teachers learn about ambitious and equitable instruction.

Shifting Concepts for Teaching

Concepts, as we mean them in the context of teacher learning, refer to teachers' narratives about given teaching practices, along with their related conditions and consequences. Rather than viewing every idea, phrase, or story as a concept, we focus on concepts that enable pedagogical judgment, the synthesized understandings that anchor and guide teachers' actions. The conceptual change

DOI: 10.4324/9781003182214-3

literature, for example, distinguishes between concepts that enable the categorization of objects or phenomena and concepts that enable interpretations about the world; the latter are inextricably linked with inferences, values, and explanations. Similarly, mathematics education research distinguishes between any beliefs teachers might hold and the specific "belief bundles" that influence teachers' shifts in goals (Aguirre & Speer, 1999).

Building on these literatures, we focus on concepts that (1) are salient to teachers in their learning environments; (2) are connected to beliefs, values, explanations, definitions, and ideologies held by teachers and communicated in their environments; and (3) enable pedagogical judgment, which we will discuss more in the next chapter. For now, we turn our attention to concepts that make possible particular interpretations of instruction, thereby facilitating teachers' actions while being grounded in their core commitments.

Because teaching and learning require collective meaning-making, we think about concepts as *narrativized actions*: the reasons and stories teachers offer to explain why they do what they do. By emphasizing narrative, we deliberately correct the overemphasis on either decontextualized practice or narrow student performance outcomes in research on teacher learning. Because ambitious and equitable mathematics instruction goes against the institutional and cultural logics of schooling, the narrative aspects of teacher learning are nontrivial, as schooling – whether accommodated or resisted – deeply shapes teachers' practice. While anthropologists of learning have theorized the relationships among individuals' thought and action, social interaction, and their cultural worlds (Erickson, 1982; Wolcott, 1982), these theories often stay at the level of the immediate environment (Eisenhart, 2021). To better grasp the cultural dimensions of conceptual change in teaching, we turn to Dorothy Holland and colleagues (1998), who explored learning-as-becoming, noting how historical legacies, social positioning, and cultural forms shape what is possible. Specifically, they described learning as personal formation, constituted by different levels of expertise, cognitive salience, and self-identification in a cultural world.

As James Stigler and James Hiebert (2009) remind us, teaching is a cultural activity, and its dominant forms in any given moment are those that best align with the institutions and values of that setting. Introducing new forms of mathematics teaching is thus a question of not merely introducing new instructional strategies, but introducing new cultural models, in which teachers may or may not grow expertise, find cognitive salience, and identify within their professional aspirations.

Thus, when teachers identify with ambitious and equitable mathematics instruction, they effectively embrace a different cultural model – one that is not aligned with prevailing expectations of schooling. Consequently, this form of teacher learning often requires that teachers explain why their practice diverges from what is taken for granted in institutional narratives and practices. Ultimately, learning to teach mathematics ambitiously and equitably requires both conceptual and cultural change. To illustrate, we next describe three conceptual and cultural shifts that ambitious and equitable mathematics instruction requires. These are meant to be illustrative, not exhaustive; we have selected shifts that anticipate our findings in Part 2.

Shift 1: How to Organize Instructional Activities

As teachers learn ambitious and equitable instruction, one major conceptual and cultural shift involves their understanding of *the nature of instructional activities*. Ambitious and equitable mathematics instruction is a *responsive pedagogy*, and, as such, builds on students' thinking and experiences. When teachers center students, their instruction moves away from extending static knowledge to students toward cultivating disciplinary learning experiences based on students' ideas (D. Cohen, 2011). Extending static knowledge – treating math as ready-made and waiting for students to master – reflects a commonplace U.S. mathematics teaching approach, described by James Stigler and James Hiebert as "learning terms and practicing procedures" (2009). This mode of instruction emphasizes memorization and algorithmic learning and de-emphasizes student sensemaking. In contrast, cultivating disciplinary learning experiences requires more complex activities like solving challenging problems and engaging mathematical practices like argumentation, modeling, and proof (Brousseau, 2006; NGA, 2010; National Research Council & Mathematics Learning Study Committee, 2001; Van den Heuvel-Panhuizen & Drijvers, 2020). Rather than having students memorize mathematical theorems and procedures, teachers provide experiences that allow students to construct mathematical meanings. In this way, ambitious and equitable mathematics instruction enables students to develop conceptual understanding and fluency alongside procedural knowledge, all the while seeing a place for their own ideas – and thus themselves – in the discipline (Gresalfi, 2009; Hand, 2012).

To meet the goals of ambitious and equitable mathematics instruction, many questions arise as teachers consider classroom activities. Teachers make decisions about the seating arrangements (e.g., rows vs. groups); how they start lessons; expectations for homework; what comprises mathematical topics and how to sequence them; what constitutes appropriate technology use; and how they communicate to students about their learning and participation. While many of these issues involve teachers' own personal conceptions of good mathematics teaching and their understanding of mathematics, these judgments are also informed by departmental or school policies, making these situated decisions (Horn & Kane, 2019). To reiterate a key idea from Chapter 1, teachers exert conceptual agency over their instructional practice, but their conceptions are also shaped by the contexts of their work.

While curriculum is certainly an important site of instructional policy,[1] experienced teachers, who are the focus of our study, often adapt or supplement curricular tasks to better align with their visions of good teaching. In this way, teachers exert conceptual agency about their teaching through their curriculum use. The research on the design and implementation of mathematical tasks offers insight into how teachers' concepts need to shift to meet the aims of ambitious and equitable instruction. For this reason, although there are many aspects of instructional activities for teachers to learn about, we focus on the conceptual shifts needed to effectively use rich mathematical tasks.

Mary Kay Stein and Margaret Smith's (2011) research has been foundational to the field's understanding of these shifts. Their work has shown how rich, cognitively demanding tasks are at the heart of ambitious and equitable instruction. Such tasks help students make important connections across mathematical concepts, procedures, and contexts. Sometimes, students apply known procedures to new situations, while other times, they invent their own strategies; both experiences allow them to deepen their mathematical understandings and build connections. Ideally, rich tasks have multiple entry points, so that everyone can get started (often referred to as a *low floor*) while also offering many avenues of exploration of core mathematical ideas (referred to as a *high ceiling*). These various avenues of exploration allow for multiple solution paths, or even multiple answers, which spur students to discuss and compare their ideas, which further deepens their understanding.

Recall the Orangey task presented in Chapter 1, which asks students to compare different ratios of orange juice concentrate and water to determine which mixture will taste most "Orangey." To solve this task, students must make sense of the context and mathematize it. Some students might orient toward the concentrate, determining that a more orangey mix has more concentrate for the same amount of water. Other students might orient more toward the water, arguing that more orangey mixes have less water for the same amount of concentrate. In other words, there are different, equally sensible places for students to start the problem and think it through. Some students might scale up mixes to a common multiple, while others might scale down mixes to one unit of water or concentrate – or they may try something else entirely. Regardless of their approach, the task encourages students to connect ideas about ratios and multiplicative relationships to the real-world context of mixing a solution. This makes the task cognitively demanding, giving students opportunities to deepen their understanding of ratio and proportion.

When teachers provide opportunities for mathematical sensemaking, they need to make tasks meaningful to students. Essentially, they must have a sense of students' knowledge and experiences in order to bridge gaps between curriculum designers' assumptions and what teachers observe firsthand. As Valerie Walkerdine (1990) described, the worlds of mathematics problems are often divorced from the quantitative realities of children's lives, leading to dissociative dissonance. She describes young, working-class girls that she observed playing a math game that involved the outlandish fiction of buying a yacht for 2 pence. The game of school mathematics asked students to put vivid, lived experiences aside (e.g., *money is a scarce resource*) while simultaneously (and perversely) insisting that mathematics allows them to "make sense" of the world.

Even in more thoughtfully designed curricula, real-world contexts are not equally accessible to all students. Rochelle Gutiérrez (2018), borrowing from multicultural literacy scholar Rudine Bishop (1990), uses the metaphor of "windows and mirrors" to describe problem contexts that allow students to see beyond their own circumstances (windows) or that reflect their lives back at them (mirrors). For ambitious instruction to be equitable, teachers must be mindful of issues of familiarity (Jackson et al., 2013). In the Orangey task, for instance, some students may not recognize orange juice concentrate but may have analogous experiences to draw on, like making lemonade from a powdered mix or chocolate milk with chocolate syrup. Changing "the Orangey task" to "the Lemony task" or "the Chocolatey task" might create a mirror that reflects students' experiences. Preserving the Orangey context, on the other hand, might create a window, allowing students to see into experiences beyond their own and, in turn, learn something new. In practice, most contexts represent windows for some students and mirrors for others; in inclusive classrooms, students encounter some windows and some mirrors, with no student having windows or mirrors all of the time.

Once teachers select or design a rich task, they need to plan how to engage students in mathematical sensemaking. This is harder than it might seem. In the QUASAR project, Mary Kay Stein and colleagues (1996) analyzed 144 mathematical tasks as they were used by different teachers. They found a tendency for teachers to maintain or lower the cognitive demands of tasks (seldom raising it), with the most cognitively demanding tasks *more likely* to have their cognitive demand lowered.[2] This trend for rich tasks to be mathematically diluted – in the researchers' words, *proceduralized* – limited students' sensemaking opportunities. When proceduralizing tasks, teachers might instruct students to use specific procedures to solve complex questions. With the Orangey task, for example, teachers might introduce the problem by telling students to divide the amount of concentrate by the amount of water. While that approach helps students arrive at correct answers – thus making the lesson smooth and ensuring that content was "covered" – it also robs students of an opportunity to solve the problem themselves and understand the ratio more deeply. A key challenge for teachers using rich tasks is prioritizing students' sensemaking, even when there are logistical pressures to do otherwise. Often, these pressures are rooted in Competing Visions of Quality (Chapter 1).

To center students' ideas and experiences, teachers need to consider several aspects of task use. First, they must ensure that the task builds on students' prior knowledge. This means that the teachers' introduction of the task – often called the launch or set-up – is highly consequential, as it helps students to understand the mathematical and contextual demands of the task, but without "overscaffolding" (as in the proceduralizing set-up described earlier). Attending to the contextual demands of the task is especially important when contexts provide windows. For the Orangey task, this might include acting out the situation (e.g., tasting different mixtures) or discussing the meaning of "concentrate" (which has a familiar and misleading homonym). Launches that attend to these issues support students' sustained engagement throughout a lesson, giving students just enough information to start without telling them exactly what approach to take (Jackson et al., 2013).

Teachers' pacing decisions are another crucial part of effectively using rich tasks. Teachers need to allow adequate time for students to explore and make sense of tasks. Marjorie Henningsen and Mary Kay Stein (2002) found that implementation problems that lower the cognitive demand of tasks arise when students have too little time for sensemaking. Additionally, teachers must make choices about having students do tasks individually, in pairs, or in groups, attending to how these arrangements support mathematical thinking. Some tasks are more *groupworthy* (E. Cohen & Lotan, 2014; Horn, 2012), encouraging different forms of mathematical thinking and sparking rich student-to-student discussion.

But having students collaborate on tasks introduces new complexities. When students drive mathematical activity in small groups, lesson quality greatly depends on interactions outside the teacher's immediate control, so they must decide how to structure peer interactions to foster student learning. For instance, in the Orangey task, teachers might format student groups deliberately: They may arrange groups around the perimeter of the classroom to show their thinking on whiteboards (Liljedahl, 2016), or they may ask groups to work at tables and create posters explaining their ideas (Horn, 2012).

Below, we use feminist philosopher of science Donna Haraway's (1988) representation of dichotomous charts to illustrate of some of the conceptual and cultural shifts that are involved in organizing instructional activities to better reflect the goals of ambitious and equitable instruction (Table 2.1). These charts do not illustrate equivalent, alternative stances. Instead, they lay bare "fixed ends of a charged dichotomy" (p. 588) through which different epistemologies, commitments, and politics flow. These lay out some of the concepts that define two cultural models of instruction. The columns thus represent onto-epistemic tensions teachers experience as they develop their conceptions about and build classroom cultures around ambitious and equitable mathematics instruction. Notably, the descriptions on the left side are better aligned with the organization of U.S. schools – the dominant cultural model. Given the Alignment Corollary described in Chapter 1, the conceptions in that column are imbued with greater institutional power, and teachers

TABLE 2.1 Core onto-epistemic tensions in conceptions of mathematics and mathematical activity in traditional vs. ambitious and equitable mathematics instruction.

Traditional Instruction	*Ambitious and Equitable Instruction*
Mathematics problems take a short time to solve	Mathematics problems must be considered in depth to be understood
Mathematics problems are solved in pre-determined ways	Mathematics problems can be approached in different ways
Mathematics is ready-made: There is nothing to figure out, just content to receive	Mathematics is in-the-making: Students can inquire into and make sense of mathematical ideas and processes
Mathematics is hierarchically organized: Students cannot progress without mastering pre-requisite learning	Mathematics is a set of connected ideas: Students can explore these connections through different forms of thinking
Mathematical tasks typically involve many similar problems that progress in difficulty	Mathematical tasks involve fewer problems, each of which fosters important connections between mathematical ideas and the world
Work completion is the most valued form of participation	Careful thinking and justification are the most highly valued forms of participation

feel conflicting obligations about whether to prepare students for schools as they are or to push back against the status quo by aligning their practice with progressive ideals (Philip, 2011).

Shift 2: What Math Class Sounds Like

Another conceptual and cultural shift required by ambitious and equitable instruction involves what math class should sound like – that is, teachers' ideas about the *nature of classroom discourse.* As classroom activities move away from extending pre-determined knowledge (like definitions and procedures) toward cultivating disciplinarily meaningful learning experiences (like problem solving), teacher-student dialogue needs to change accordingly (D. Cohen, 2011).

Whole-Class Discourse

In traditional instruction, teachers' ideas control whole-class discourse, making lessons predictable – but also foreclosing students' sensemaking opportunities. Typical whole-class discourse is formatted with teachers *initiating* questions; students *responding*; and teachers *evaluating* their responses. This I-R-E pattern (Mehan, 1979) focuses dialogue on *known-information questions,* allowing the teacher to control the flow of interaction, thereby reducing the uncertainty of how lessons progress. Importantly, when lessons are more predictable, they are more likely to be smooth – which, as we discussed in Chapter 1, is often taken as an important (albeit limited) indicator of teaching success.

In the Orangey task, for instance, teachers might ask the question, "How can we write the ratio of Mix B?" There are limited possible student responses and there are clear right and wrong answers, allowing teachers to respond evaluatively (e.g., "That's right!", "Anybody else?"). However, when teachers shift their instructional goals toward student sensemaking, their questions also shift toward *information-seeking questions*, queries to which they genuinely do not know the answers. Often, these questions require students to reflect on their own thinking, like, "How did you decide which mix was orangier?" The shift from known-information questions to information-seeking questions is crucial for inviting student sensemaking, because the latter question format promotes students' reflection and articulation of their own ideas.

Once again, the move toward ambitious and equitable instruction invites ambiguity into the classroom: As discourse becomes less teacher-centered, there are more in-the-moment interpretive demands. Information-seeking questions require different forms of teacher listening and on-the-fly decision-making (Hintz & Tyson, 2015). Additionally, information-seeking questions introduce tensions about who controls classroom interactions, since students have more opportunities to introduce unexpected ideas, many of which may seem "off topic" (Emanuelsson & Sahlström, 2008). For these reasons, teachers require considerable support in making this change as they manage student-centered whole-class discourse (Michaels et al., 2008).

While much classroom discourse is anchored in questions, these questions – no matter their format – create longer sequences that differently support student sensemaking (Lefstein et al., 2015). For instance, teachers might ask information-seeking questions, like "How did you decide which mix was orangier?" to individual students one after another, without building connections across their responses. While the question format invites student thinking, the overall discourse sequence does not support much sensemaking; instead, it becomes a show-and-tell of solutions, with no explicit press toward mathematical interpretation. Alternatively, if teachers juxtapose one group's approach (e.g., dividing the amount of water by the amount of concentrate) with another group's approach (e.g., inversely dividing the amount of concentrate by the amount of water), they bring in more mathematical richness. Comparative discussions – questions like, "How are these approaches different?" – require students to listen to each other's ideas and use mathematical practices like argumentation and representation as they sort through them.

With this additional richness comes myriad facilitation issues for teachers. In comparing solutions to the Orangey task, for example, students might notice that the "orangiest mix" has the smallest number in the water-to-concentrate approach and the biggest number in the concentrate-to-water approach. Are they both reasonable approaches? Which ratio better communicates "orangeness?" As students consider these issues, teachers must decide if exploring reciprocal ratios is a productive avenue of discussion, which is not a straightforward judgment. Instead, teachers weigh multiple, competing "goods," balancing judgments about whether students grasp the mathematical issues at hand, which students have had opportunities to present ideas recently, if there is enough class time for this exploration, and whether the topic is worth the effort expended in getting everybody to understand the debate. (See Lampert, 1990, 2001 for vivid first-person descriptions of these tensions.)

Indeed, from a mathematical learning perspective, invitations to compare solutions create high press for students' understanding (Kazemi & Stipek, 2009) and support students' concept development (Rittle-Johnson et al., 2020). In such discussions, students engage in key mathematical practices; they make conjectures and critique each other's reasoning. Over time, students learn to expect such discourse practices, constituting sociomathematical norms for the classroom (Yackel & Cobb, 1996). In this way, teachers shape productive mathematical discourse at multiple timescales – within questions, across talk sequences, and over multiple lessons – eliciting and building on students' ideas. In turn, students become acculturated to such discussion formats, changing their own understanding of what it means to do mathematics (Gresalfi, 2009).

At slightly broader scales, sociomathematical norms can become a part of a departmental culture, as was the case at Railside School (Nasir et al., 2014). A whole set of resources emerged around the onto-epistemics of doing mathematics and being mathematical learners that aligned with ambitious and equitable instruction: curriculum, language, teaching practices, narratives about students, and students' own expectations for themselves (Boaler & Staples, 2008; Horn, 2010; Nasir et al., 2014). This, in turn, accelerated mathematics teachers' development of ambitious and equitable instructional practices, as the institution became more aligned with these onto-epistemic commitments. We take the case of Railside as an existence proof of the situative, distributed nature of concepts for ambitious and equitable mathematics instruction: The department became a cultural world where this was the norm, shifting many institutional forces toward the ideals of this form of practice.

Small-Group Discourse

Undoubtedly, centering student thinking in whole-class discourse introduces new uncertainty for teachers, threatening the smoothness of lessons. Handing off the mathematics to students in small groups brings about even more. Even with the most intentional modeling of productive mathematical talk in whole-class settings, students do not necessarily succeed in conducting such discussions with their peers. Putting students to work together, outside teachers' immediate oversight, grants students greater agency but also opens the door for their social worlds to more fully enter the classroom, in all their colorful complexity. Beyond sharing their mathematical thinking, students' small-group conversations may bring out their fuller selves – an opportunity that is an important inclusion strategy (Davidson, 1996; Joseph et al., 2019). At the same time, they bring in friendships, romances, rivalries, prejudices, hobbies, and interests (Esmonde & Langer-Osuna, 2013; Langer-Osuna, 2011), opening up new facilitation challenges for teachers.

Student *status* (E. Cohen & Lotan, 2014) is a useful concept for supporting productive student collaboration. Status is a core concept in sociology, a constant fact of social life. In classrooms, students' academic and social status are salient, and those with high status are seen as more desirable – smarter (high academic status) or more popular (high social status). Students with low status are seen as less desirable – not smart (low academic status) or less popular (low social status). These two dimensions of status create complicated interpersonal dynamics for teachers to interpret and respond to. When status issues are prevalent, high-status students typically dominate groupwork, doing most of the talking – and thus most of the sensemaking – with their peers. Low-status students are often peripheral (or, worse, marginalized) in groupwork, either overtly (because no one listens to their ideas) or subtly (because no one expects or asks them to contribute). Mitigating status differences in small-group discourse is crucial for supporting meaningful and equitable engagement, and this poses numerous instructional puzzles for teachers. These challenges are exacerbated when teachers are invested in the onto-epistemics of mathematical competence as synonymous with quick and accurate calculation (Boaler, 1997; Horn, 2007). As we elaborate in the next section, such a narrow view of mathematical ability reinforces hierarchies in student status that naturalize the idea that only some students are worth listening to.

Fostering productive small-group discourse points back to questions about organizing instructional activities. Since collaboration is not the norm in many schools, teachers often need to help students learn how to work together effectively in ways that address the social reality of status issues. In mathematics classrooms, students themselves may be invested in narrow views of mathematical competence, only adding to the complexity. In other words, teachers are not the only ones who need to shift their onto-epistemic conceptions of mathematical ability: They must also coax students to develop new understandings of what being mathematical means. Such practices are examples of the cultural change required for ambitious and equitable instruction.

Instructional strategies that facilitate meaningful mathematical learning and address widespread notions of competence exist. For instance, in their Complex Instruction framework, Elizabeth Cohen and Rachel Lotan (2014) suggest disrupting academic status by using groupworthy tasks that require multiple abilities. Group roles can also help students collaborate and overcome negative status dynamics: A student in the role of Recorder/Reporter ensures the group's ideas are written down, while the Resource Manager collects materials for the task and finds additional resources (including the teacher) when the group needs help. As students develop ways of working together, using these scaffolds, teachers constantly rely on their own ideas about productive collaboration, interpreting, and monitoring students' participation, drawing on structures and prompts that keep students on track (Ehrenfeld & Horn, 2020).

In addition to facilitating discourse and collaboration in small groups, teachers aiming for ambitious and equitable instruction need to manage transitions between small-group and whole-class discussions. For instance, they may want students to share their group's thinking with the whole class. How do they decide when such discussions should happen, given groups' different paces? Calling the class together too soon can interrupt important thinking in progress. Waiting too long might lead some students to sit idly as their classmates finish. What are useful structures for sharing thinking? How do they ensure that students actively listen to each other? We spell out these decision points to highlight the complexity and centrality of teachers' judgments in realizing ambitious and equitable instruction. Once again, we represent the shifts in what mathematics class sounds like through a dichotomous chart, emphasizing the continuous nature of these conceptions and the tensions that flow between them (Table 2.2).

TABLE 2.2 Core onto-epistemic tensions in conceptions of organizing classroom discourse.

Traditional Instruction	Ambitious and Equitable Instruction
Teachers effectively present ideas to convey mathematics to students	Teachers design effective learning environments to convey mathematics to students
Teachers attend to correct answers	Teachers attend to student thinking and help students engage with each other's thinking
Students are passive recipients of knowledge	Students' confusion and disagreements are expected and become the basis for instruction
Learning is best achieved through individual effort	Learning is best achieved through a combination of individual and collective effort
Participation is optional	Participation is expected of every student, regardless of prior achievement

Shift 3: Who Belongs in Math Class

Given school mathematics' role in sorting and ranking students, mathematics classrooms are often marked by a *culture of exclusion* (Louie, 2017), the restrictive culture where only some students are seen as deserving of mathematical learning. Sorting and ranking are facilitated by the previously described narrow ideas about mathematical smartness, conflating it with quick and accurate calculation (Boaler, 1997). Indeed, many mathematics classrooms are organized hierarchically, with some students seen as "high" or "fast" – or simply "math kids." Others, in contrast, are seen as "low" or "slow" (Horn, 2007) – thus not really belonging in the class, by both their own and others' judgments. These conditions lead to another crucial conceptual and cultural shift for teachers around *the nature of mathematical competence.*

In fact, this shift is fundamental to the whole enterprise of ambitious and equitable instruction. Teachers' successful use of rich tasks and fostering of mathematical discourse depends on seeing students as worthwhile mathematical thinkers. To authentically pursue students' ideas, teachers must view many different ways of thinking as contributing to conversations. Broadening what it means to be mathematically smart – and making this shift legible to students – is, in part, a question of classroom design (Horn & Gresalfi, 2021), requiring teachers to have deep understandings of those competencies and how to foster them.

Importantly, expanding mathematical competence does not mean watering down mathematics. Indeed, the opposite is true: Expanding mathematical competence means rendering authentic mathematical smartnesses both visible and consequential in classrooms. Looking at mathematics as a field, we see that its great accomplishments have not come about from quick and accurate calculations, but from other kinds of insights, creativity, and intelligence: asking good questions, making astute connections, working systematically, seeing patterns, illustrating representations, and so on (Horn, 2012, 2017).

To expand notions of mathematical competence, teachers must also reject deficit narratives of students, their families, and their communities. Stereotypes about who does mathematics are informed by broader cultural narratives about race (Chen & Buell, 2018; McGee & Martin, 2011; Shah, 2017), gender (Leyva, 2017; Walkerdine, 1998), social class (Boaler, 2002; Walkerdine, 1998), language (de Araujo et al., 2018; R. Gutiérrez, 2002), disability (Lambert, 2015; Lewis, 2014), and their various intersections in students' identities and lives (Bullock, 2018; Gholson & Martin, 2019; Joseph, 2020; Leyva & Alley, 2021).

Learning to replace deficit narratives of students with asset-based ones also shapes how teachers address the status perceptions described in the previous section. A key conceptual refinement for teachers is the distinction between *doing school* and *doing math.* The former invokes students' compliance and conformity to the norms of schooling, which, in the U.S., are largely shaped by white, middle-class, heterosexual, monolingual English-speaking culture (Battey & Leyva, 2016; Pascoe, 2011; Valenzuela, 1999). Doing school – turning in homework, being quiet in class, using "acceptable" language and grammar – is not the same as doing math. Yet many decisions about students' mathematical learning opportunities, including enrichment programs or higher-level math placements, depend on students' adherence to schooling norms (Calarco et al., 2022; Horn, 2018). By disentangling students' compliance and assimilation from their mathematical ability – and especially by embracing multiple mathematical abilities – teachers make space for more students' authentic participation in their classrooms.

Gloria Ladson-Billings (2014) has characterized this disentangling of doing school and doing math as part of a larger shift toward culturally relevant pedagogy. In her framework, she describes how culturally relevant teachers focus on students' academic success, in part, by making the hidden curriculum of schooling explicit. Her framework also encompasses other issues we have touched on. Culturally relevant teachers respect students' cultures and use them as meaningful vehicles for learning by letting students use their home languages or by including tasks that offer mirrors to reflect students' lives. Furthermore, culturally relevant teachers develop critical consciousness about the inequities baked into the educational system and society – they question the status-quo understandings represented on the left sides of the dichotomous charts in this chapter. This may show up in their curricular choices (Gutstein, 2006) or in their advocacy for students, which may include what Rochelle Gutiérrez (2016) calls creative insubordination. Gutiérrez describes creative insubordination as including acts like decentering the achievement gap, questioning what mathematics is taught in school, highlighting the humanity and uncertainty in mathematics, positioning students as mathematical thinkers, and challenging deficit narratives about historically marginalized students. As Ladson-Billings said in a recent interview:

> A hallmark for me of a culturally relevant teacher is someone who understands that we're operating in a fundamentally inequitable system – they take that as a given. And that the teacher's role is not merely to help kids fit into an unfair system, but rather to give them the skills, the knowledge, and the dispositions to change the inequity. The idea is not to get more people at the top of an unfair pyramid; the idea is to say the pyramid is the wrong structure.
>
> *(Fay, 2019)*

In narrating three conceptual shifts underlying ambitious and equitable mathematics instruction, we treat them with equal weight. However, as teacher educators, we find that broadening notions of mathematical competence is a key lever for developing critical consciousness. Once teachers begin to question why the quick and accurate calculation is the most prized form of mathematical smartness, it opens up larger, systemic questions about who counts as smart and why (or why not), what it means to do school, and the very nature of mathematics. Broadening notions of mathematical competence challenges status issues and also invites more students' ideas into classroom conversations. This, in turn, fosters a sense of belongingness, which has repeatedly been shown to be a core dimension of equitable teaching. *Belongingness* refers to people's innate need to establish close relationships with others. It develops when members of a classroom community create authentic connections through pleasant, meaningful interactions. Belongingness is important for all students, but it is especially important for those who do not fit the stereotypical image of a "math person" – in the U.S., this is typically girls and non-binary students, students with disabilities, Black, Indigenous, and Latinx students, students living in poverty, as well as students who have strong interests in the arts or humanities. Building off a broad sense of mathematical competence, teachers committed to ambitious and equitable mathematics teaching must develop asset-oriented ways of thinking about their students, expanding notions of what it means to do mathematics and finding ways to connect with the diverse learners in their classroom.

Fostering belongingness is critical to ambitious and equitable mathematics instruction. When teachers ask students to share their "rough-draft thinking" (Jansen et al., 2017) – ideas that may not be fully formulated yet – they are asking students to make themselves vulnerable, to reveal ideas that they are not confident in. This socially risky activity can be mitigated by a sense of belongingness: A classroom culture of warmth, humor, and acceptance assures students that their emerging (and possibly incorrect) understandings will be received with generosity and care (Horn, 2017). Without such assurance, students may remain incalcitrant in the face of teachers' bids for them to share, which dooms student-centered instruction before it can even begin.

The relationship-building needed to foster belongingness is likely the most situated of these conceptual shifts. Classes – and individual students – have different personalities. Teachers' own social identities and "historicized selves" (Chen, 2020) come into play as they develop relationships with their students. Teachers' racial, linguistic, academic, class, gender, religious, and other identities invite different kinds of connections with students, along with their personalities, preferences, and temperaments.

Representing the shifts in who belongs in math class through a dichotomous chart, we aim to summarize the tensions that teachers negotiate as they aim for ambitious and equitable teaching (Table 2.3).

Certainly, many of the ideas reviewed in this chapter are not new. In the U.S., supporting students to engage in disciplinary practices of mathematics first gained popularity during the New Math era of the 1960s. By the 1980s, the National Council of Teachers of Mathematics (NCTM) began synthesizing research on ambitious teaching into influential policy documents, sparking funding agencies and curriculum developers to invest in their vision of good mathematics teaching. These efforts led to professional conferences, publications, and professional development workshops, as NCTM continued to advocate for teaching students mathematics for understanding. Similarly, curricula built around rich tasks have been available for decades. Targeting the inclusiveness required for ambitious and equitable mathematics instruction,

TABLE 2.3 Core onto-epistemic tensions in conceptions of who belongs in mathematics classrooms.

Traditional Instruction	*Ambitious and Equitable Instruction*
Students' individual understandings and identities do not influence instruction	Students' individual and collective understandings, along with their broader identities and experiences, shape instruction
Quick and accurate calculation is the most important form of mathematical smartness	Multiple forms of mathematical smartness are valued and cultivated
Students wait for the "smart" student to answer questions correctly	All students' ideas are valued for the class' collective sensemaking, and "incorrect" answers can be important contributions to a shared understanding
Status strongly influences student participation	Student participation is broadened by fostering equal-status interactions
Assessment tools make definitive judgments about student achievement, potential, and smartness	Assessment tools encourage persistence and learning by offering opportunities to revise

organizations like TODOS and the Algebra Project have advocated for and developed culturally responsive and social justice-oriented mathematics tasks. Despite this long history and dedicated focus, ambitious and equitable instruction is not typical in most U.S. mathematics classrooms. National surveys show that teachers feel especially underprepared to build inclusive instruction for historically marginalized learners and do not receive adequate professional development to meet this goal (NASEM, 2020), contributing to the persistence of traditional instruction.

Conceptual Infrastructure for Teachers' Understandings

Saying that teachers need to make conceptual shifts to enact ambitious and equitable mathematics instruction emphasizes concepts in individuals. In this section, we consider how conceptual systems are distributed across people and contexts – especially since ambitious and equitable mathematics instruction is best fostered within teacher communities (Boaler, 2002; Nasir et al., 2014). We consider how concepts extend across people, groups, activities, and settings, as meanings are created, reinstated, and negotiated through interactions and over time. In an investigation of workplace learning, Rogers Hall and Ilana Horn (2012) state:

> We understand *concepts* as recurring patterns of purposeful activity that are *distributed* over people and technologies in work practice. Related to this, learning is an active process of *distributing* cognition over people and things. Analyzing the work of concept formation thus requires tracing

how these distributions are accomplished. [...] Because concepts in our framework are distributed over patterns of activity and technologies, they are integral to the representational infrastructure of work.

(p. 241)

To be sure, the notion of concepts extending beyond individuals and into cultural practices is a widely accepted premise in the sociocultural tradition (Rogoff et al., 2003; Rosebery et al., 2010; Saxe, 2012). As Hall and Horn suggest, this distribution can be investigated by looking at particular institutional organizations, cultural practices, and representations – what we call *conceptual infrastructure* – that shape teachers' actions and narratives about mathematics teaching (see also Hall & Jurow, 2015). For this reason, we focus on how teachers draw on conceptual infrastructures to guide their instructional action and interpretations thereof.

Teachers' collegial conversations offer a fruitful site for identifying teachers' use of conceptual resources and how they are distributed over time; we treat these conversations as found talk-aloud protocols for teachers' sensemaking (Horn, 2013). When we examine these conversations, we look for teachers' pedagogical reasoning in their talk – as they plan lessons, debrief their day, or consult with colleagues – and we analyze the resulting narratives to make sense of their thinking about their work. The narratives have all the features of concepts, as described in the conceptual change literature (diSessa, 2006): They have structure; they identify critical attributes of the underlying phenomena they seek to explain; they posit relationships and processes. For instance, a teacher might narrate students as not learning because they did not pay attention; another teacher might narrate a lesson as going poorly because they used unfamiliar materials. In both instances, these narratives offer interpretations of instructional moments, with assertions about the underlying causes of these events. These resources, and the ways they inform teachers' sensemaking, reflect the subject-world relations discussed in Chapter 1, highlighting both the conceptual embeddedness and possibility for conceptual agency in teacher knowledge.

In this section, we offer a brief overview of some key conceptual resources we have identified: the structure and rituals of schooling, representations of teaching practice, onto-epistemic stances, activity structures, and problem frames. We also offer a caveat similar to the one in the previous section: Although we disaggregate these resources for analytic clarity, they work dynamically together to help teachers interpret and understand their teaching.

Structures and Rituals of Schooling

Schooling is replete with structures and rituals that have become so familiar as to be taken for granted: One adult leading many children; time carved into discrete units of activity devoted to different subjects; school bells marking transitions between activities; homework; testing; and so on. None of these is essential to learning, but they are emblematic of what we think of as schooling. This social organization conveys what institutional theorists refer to as institutional logics (Friedland & Alford, 1991; Thornton & Ocasio, 2008) – the broad, socially constructed practices and rule structures that shape institutional reality, forming a kind of common sense.

These structures and rituals serve as a backdrop for teachers' sensemaking, a hidden curriculum that lends schooling its unique grammar (Giroux & Penna, 1979; Tyack & Tobin, 1994). For teachers, schooling itself undergirds a common sense about practice (Horn & Kane, 2019). That is, we hear teachers explain their actions by invoking extant practices: Students must study to help them learn to study; teachers assign homework to develop students' homework habits. Although this kind of reasoning is, on one level, tautological, it also points to cultural repertoires of practice, the archetypical actions that make what teachers do recognizable as teaching (K. Gutiérrez & Rogoff, 2003). Mathematics classrooms also have rituals around textbooks that create another kind of common sense about instructional activities. In U.S. classrooms, teachers often defer to textbooks as the ultimate mathematical authority, following their lead – the sequence of topics, the representations of ideas – as a definitive way to present mathematics (Herbel-Eisenmann, 2007).

Onto-Epistemic Stances

The repertoires of schooling practices that shape instruction take on different meanings depending on their enactment. Teachers may insist that students study because they need to learn to study, but what that studying looks like and how it is positioned as a cultural practice offers different meanings. Teachers may tell students that they must study to show they are serious about their education; in contrast, teachers may tell students that they must study because that is how to play the game of school. Studying, as an ontological and epistemological entity, gets communicated in different ways. We refer to these ways of understanding what studying – or any other practice – involves as *onto-epistemic stances* (Horn & Kane, 2015; Warren et al., 2020). Onto-epistemic stances communicate what something is, how it can be known, and whether it is of value.[3]

Teachers' onto-epistemic stances are visible when they talk to colleagues, who, in turn, become important sources of teachers' sensemaking about their instruction (Coburn, 2001; Horn & Little, 2010). Analytically, teachers' conversations reveal onto-epistemic stances on different timescales. At the smallest timescale, single turns of talk manifest these stances, for instance, "What matters here is getting the kids to share their thinking." We view such an utterance as an *onto-epistemic claim* since it communicates that (a) kids can share their thinking, (b) this practice applies to the current discussion, and (c) helping kids share their thinking is valuable. At a slightly broader timescale, questions imply onto-epistemic stances (Horn & Little, 2010). For instance, the question, "Do you think the kids fell apart because it was the end of the day?" communicates a stance that fatigue is something students experience and that it may negatively influence student behavior. Over longer time periods, onto-epistemic stances surface through activities, as interactional emphases and attention reveal commitments to what can be known and how to know it, while providing different interpretive resources on the work. These can be institutional activities, like placing students in math classes, or they can be local to classrooms, like grading student work. In all of these examples, we see concepts being conveyed. They propose narrative action, imply cause and effect, and offer explanations.

Representations of Practice

We also have seen teachers' concepts revealed in their conversations through *representations of practice* (Little, 2003). Teachers typically discuss their work asynchronously from active instruction (the Asynchronicity of Reflection Premise), so when they talk about lessons – whether consulting with colleagues or participating in professional development – they need to reconstruct critical aspects of teaching to ground these discussions. Representations of practice anchor teachers' shared meanings – in Vygotskian terms, their inter-subjectivity. Sometimes, curricular materials or lesson plans serve as important representations of practice. Other times, these representations are discursively constructed. Two common forms of teacher talk that reconstruct classroom events for joint consideration are *replays* and *rehearsals* (Horn, 2010). Replays provide blow-by-blow accounts of classroom interactions or ongoing events, while rehearsals allow teachers to act out interactions in a more general or anticipatory fashion. Replays and rehearsals are common ways for teachers to share ideas about classroom discourse, as they import simulated classroom dialogue for their colleagues' consideration. For instance, a teacher might "rehearse" a set-up for the Orangey task by telling colleagues how she typically introduces it:

> First, I ask the kids if they've ever had drinks from a mix. I get some hands, and a kid will usually say something like, "My grandma gives me Kool-Aid!" or whatever. Then I ask if anybody has ever heard of *orange juice concentrate*.

Such enactments allow teachers to share ideas about instruction, modeling the importance of bridging the Orangey task with students' experiences. Along with the collegial support, conversations like this also coordinate teachers' onto-epistemic stances about what it means to teach. The metamessage of this kind of communication is that *tasks can be launched by linking the context to students' experiences,* contributing to concepts in circulation in that setting.

Activity Structures

Teachers reveal their conceptions of teaching through activity structures, the patterned ways tasks get carried out in the course of their work, whether individually or with colleagues. For instance, lesson planning may be done as an individual or group activity, a variation that communicates different notions of who is responsible for designing instructional activities. To qualify as activity structures, tasks may be formalized or improvised as people work, or some combination thereof. Over time, repeated activities – department meetings, lunchroom talk, lesson planning, reviewing student work – may become routine in that they are predictably patterned and recurrent (Horn & Little, 2010).

In teachers' conversations, activity structures and representations often work together to inform their conceptions of different aspects of instruction. For example, in the activity of looking at student work, teachers might look at the same samples of work or at their own class sets. These differences in the activity require different representations (a shared set of work vs. individual sets of work), and this shapes the discussion. In the former case, they can delve into a subset of solutions to investigate the same students' thinking together. In the latter case, teachers' firsthand knowledge of students can be brought to their interpretations of student solutions.

Problem Frames

Teachers also reveal their concepts for teaching through problem frames. Closely related to activity structures, problem frames describe how issues are defined through interactions (Goffman, 1974). Frames organize teachers' collective attention, shifting the meaning of activities (Horn, 2007), differentially positioning teachers as agents in problems they face (Bannister, 2015), changing the concepts they have access to.

To understand the connection between frames and onto-epistemic stances, consider again the activity of looking at student work. Teachers might frame this activity as "aligning grading standards," posing the question: *Do we all agree what good work looks like?* Alternatively, teachers might frame the activity as "making sense of students' thinking about key mathematical ideas," posing the question: *How do students understand this concept?* The first frame might support discussions about scoring criteria, while the second might lead to identifying student understandings. Both frames can support teachers' concept development through the "same" activity, but they foreground different issues.

Problem framing is consequential to teachers' interpretations of problems of practice. Problem frames provide the context clues people use to make sense of uncertainties in their environment. Linguist George Lakoff (2014) has shown how frames, at a societal level, render different kinds of solutions as "sensible;" their internal metaphors communicate logics that can be taken-for-granted as common sense. This phenomenon also relates to the well-known observation in cognitive psychology that context clues shape – or even distort – people's interpretations. For instance, in the Müller-Lyer illusion (Lewis, 1908), two lines of identical length are seen as shorter or longer by the addition of fins that point either inward or outward. Problem frames are similarly influential to perception and sensemaking: They are the fins that make one reality look more viable than another. In this way, problem framing is important for fostering teachers' conceptual agency, as it shapes their interpretations of instruction.

Ultimately, structures and rituals of schooling, epistemic stances, representations of practice, conversational frames, and activity structures form a conceptual infrastructure that shapes how teachers make sense of teaching. Identifying the conceptual resources teachers draw on is important for supporting teachers' concept development, as they offer potential levers for transforming teachers' understandings toward ambitious and equitable instruction – ones that live outside individual teachers and extend to schools as workplaces. Indeed, by focusing on conceptual infrastructure and the sensemaking resources therein, we can imagine ways to re-mediate teaching environments to introduce new ideas about teaching and learning and give them social power and salience (Jurow et al., 2019).

Developing Integrated Concepts About Teaching

Thus far, we have outlined the way concepts about core aspects of instruction shift as teachers develop ambitious and equitable mathematics instruction. Then, aligned with our situative view of teacher learning, we described the conceptual infrastructure and resources within it that extend beyond individual teachers into their schools and workplaces, shaping their interpretations of practice.

As we alluded to in the first section, the conceptual shifts we described are closely interrelated. Using rich tasks becomes a way to support students' sensemaking, which happens through classroom discourse; meaningfully implementing tasks and fostering a sense of belongingness are likewise interdependent, as students often enter math class with a fear of not knowing what to do when presented with complex problems. Similarly, supporting productive mathematical discourse requires teachers to elicit students' thinking, which students will likely only share if they experience a sense of belonging in the classroom, making these mutually reinforcing.

How do concepts for teaching change as teachers become more accomplished in ambitious and equitable mathematics instruction? In a cross-sectional comparison of teacher workgroups' pedagogical reasoning based on their instructional practice, we found several notable differences in the quality of their conceptions (Horn & Kane, 2015). In addition to spending more time discussing problems of practice and more consistently considering students' perspectives than teachers in the comparison groups, the most accomplished teachers employed what we called *ecological views of instruction* to interpret problems of practice. That is, they constantly considered interrelations among various aspects of teaching, especially the nexus between teaching, students, and mathematics. If an accomplished teacher was using the Orangey task, for instance, she might consider the mathematics in light of what students had previously learned about ratios; she might consider adapting or modeling the problem context to ensure students understood it; and she might think about students working in pairs or groups, depending on her interpretations of classroom dynamics. In contrast, teachers less accomplished in ambitious and equitable instruction might only plan an engaging launch or discuss issues of pacing, without considering the interrelationships among questions of pacing, engagement, and student understanding.

We are certainly not the first to observe that accomplished teachers develop integrated views of instruction as their teaching becomes more responsive to students' thinking and experiences. For instance, Barbara Levin and Paul Ammon (1992) noted teachers developed multidimensional forms of pedagogical thinking as they learned more student-centered instructional practices, which the authors described as a "trend toward differentiation and integration" (p. 25). In her detailed, first-person account of her own mathematics teaching, Magdalene Lampert (2001) illustrated specific classroom moments that drew on such integrated knowledge. Similarly, Fred Korthagen (2010) explained how teachers form *gestalts* about prototypical teaching situations. He described gestalts as holistic, mid-level theories about instruction, encompassing rational thought, experience, and emotion, that help them interpret and act on classroom situations. Likewise, Adam Lefstein and Julia Snell (2013) described teachers learning to teach dialogically – with rich classroom discourse – which involved honing connections across their sensitivity-repertoire-interpretation-judgment, suggesting a tight coupling between judgment and practice. Finally, Anne Edwards (2010) used a Vygotskian lens to describe how teachers come to view teaching as a "resourceful practice." This means that, as teachers develop expertise, they learn to interpret and act on their environments more effectively, often by drawing on resources like material artifacts, colleagues, mentors, and students themselves.

These images of integrated, holistic, and resourceful teaching echo Paulo Freire's (1970/2018) notion of *praxis*, the "reflection and action directed at structures to be transformed" (p. 126); notably, praxis points to the need for cultural change as a part of integrated understandings of practice. Teachers' increased capacity to connect purpose and action – the adjustment and re-adjustment of action to better align with goals, and, at least for Freire, to change institutional structures – is captured by numerous close observers of teacher development. Returning to our teacher learning framework, we take this as showing that the telos of teacher learning of ambitious and equitable instruction is not simply shifting concepts about teaching; rather, it also involves developing ecological, holistic, integrated views of these aspects of teaching and continually exploring their interdependence in different situations.

Conceptual – and Cultural – Change

As we describe how teachers' concepts develop as they learn ambitious and equitable mathematics instruction, it is worth reiterating that the related images of good teaching are uncommon in U.S. schools. As teachers navigate Competing Visions of Quality and contend with the Alignment of instruction to typical schooling processes, they have to not only learn how to teach differently, but also bring about cultural change to legitimate their instructional practice to other stakeholders, including students, parents, colleagues, and administrators – Freire's praxis.

The extent to which teachers align with institutional norms has consequences for their development. Of course, it is unreasonable to expect people to be in full alignment on questions of education. With some shared values, different notions of how to achieve them, and trust, stakeholders can discuss ideas and learn from one another – a situation characterized by productive friction (Ward et al., 2011) where the work of reconciliation supports conceptual clarity. It is altogether another situation when ideas about teaching become incommensurate, making it difficult to find common ground and requiring another set of practices entirely (Sfard, 2019). Alignment, friction, and incommensurability between teachers' commitments and institutional logics implicate the effort required to learn and use ambitious and equitable instruction.

For the most part, U.S. schools remain misaligned with the goals and practices of ambitious and equitable instruction. Why is this? Returning to the shifts outlined earlier, we note an important theme: Ambitious and equitable instruction, captured in the onto–epistemic commitments represented on the right side of Tables 2.1–2.3, increases uncertainty and creates a more complex terrain for teachers' interpretations and actions. Consider the example of students collaborating in small groups. Teachers relinquish some control over discussions, opening spaces that may (or may not) be used for math talk, making it harder to ensure inclusion and to predict the pace of lessons. Even when students collaborate well, they may introduce unexpected questions or unforeseen solutions, and teachers may not know how to respond. These outcomes complicate planning, since students' unanticipated actions require teachers to make on-the-fly interpretations and adjustments, potentially deviating from what they originally planned. The improvisational nature (Philip, 2019) of ambitious and equitable mathematics instruction comes from its responsiveness. As such, though there are general principles and guidelines for what it looks like, it cannot be scripted or standardized. Naturalizing uncertainty goes against the dominant culture of schooling, the onto-epistemic commitments represented on the left side of Tables 2.1–2.3.

A notable change in 21st-century U.S. schools is the move toward standardization through standardized tests and state standards that dictate what topics should be taught and when. Although standards and accountability were originally aimed to make content goals more transparent and consistent, they have ultimately become a form of technocratic control (Mehta, 2013). As a result, mathematics teachers feel tremendous pressure to keep up with pacing guides, which makes it difficult to incorporate rich tasks that take too much time. Additionally, standardized tests work against broader notions of mathematical competence that increase participation in mathematics classrooms. They are typically norm-referenced and thus designed for ranking and sorting. They define narrow student learning goals – for instance, requiring that students add fractions by finding common denominators, without needing to understand the concepts behind this strategy (Garner et al., 2017). Since the early 2000s, widespread accountability policies have exacerbated this trend, with schools facing sanctions for not meeting performance targets. These conditions are especially harsh in urban schools that serve historically and currently marginalized communities (Au, 2007; Horn, 2018).

Aside from standardizing mathematics education, other structural elements of U.S. secondary schools pose substantial challenges to ambitious and equitable instruction. Again, the routines and structures of schooling naturalize institutional logics that convey common-sense concepts of teaching and learning. For instance, the bell schedule divides the school day into multiple class periods that limit time available for rich tasks. Overcrowded classrooms impede teachers' capacity to listen to all of their students and understand their thinking. Ranking and sorting individual students challenges the goals of collaboration and instead exacerbates status issues. As our empirical chapters show, teachers in our study faced challenges around these structural realities. Narrativizing their teaching practice, however, supported teachers in identifying and sometimes managing tensions between their aspirations and institutional logics, in part by offering them opportunities to connect their pedagogical actions, reasoning, and responsibilities.

Summary

In this chapter, we presented the second aspect of our theory of teacher learning – the *telos* of developing concepts at the core of ambitious and equitable instruction. We summarized prior research on ambitious and equitable mathematics instruction to describe three crucial, highly interrelated shifts about teaching. Regarding instructional activities, teachers move away from decontextualized mathematics problems toward designing and using rich tasks. Regarding what mathematics class sounds like, teachers move from teacher-led, answer-driven discourse toward student-centered, sensemaking discourse. Finally, in terms of who belongs in mathematics classes, teachers shift from the dominant culture of exclusion that ranks and sorts students according to narrow notions of mathematical competence toward inclusive classrooms that value multiple forms of mathematical smartness and embrace students' experiences, cultures, languages, and interests. These conceptual shifts also require cultural shifts, adding another layer to this undertaking.

Alongside our situative view of learning, we argue that these conceptions do not entirely reside within individual teachers. Instead, they extend across structures and rituals of schooling, onto-epistemic stances, representations of practice, activity structures, and problem frames that teachers draw on as they make sense of their work – all of which function as conceptual resources for their pedagogical reasoning. Finally, in addition to describing the conceptual shifts that undergird teachers' development of ambitious and equitable instructional practice, we note a strong research consensus that teachers' concepts become more integrated as they develop responsive and conceptually-driven instruction, developing ecological thinking about teaching.

Situativity, as a perspective, gives us additional resources to leverage in supporting teachers' development toward ambitious and equitable mathematics instruction. It also underscores the significant institutional challenges teachers confront, as schooling arrangements are not always hospitable to these practices.

Notes

1　See Hirsch, 2007, for an overview of many of the standards-based curricula developed in the U.S. to align with goals for ambitious and equitable instruction.
2　We see this as more evidence for the Smoothness Corollary in Chapter 1.
3　In earlier writing (Hall & Horn, 2012; Horn & Kane, 2015) we have referred to these as epistemic stances. As we argued in Horn and Kane (2019), because concepts in teaching often exist in an ideological (rather than physical) realm, teachers' statements and actions often extend beyond the knowledge claims of epistemology into the ontological claims about what something actually is.

References

Aguirre, J., & Speer, N. M. (1999). Examining the relationship between beliefs and goals in teacher practice. *The Journal of Mathematical Behavior, 18*(3), 327–356. https://doi.org/10.1016/S0732-3123(99)00034-6

Au, W. (2007). High-stakes testing and curricular control: A qualitative metasynthesis. *Educational Researcher, 36*(5), 258–267.

Bannister, N. A. (2015). Reframing practice: Teacher learning through interactions in a collaborative group. *Journal of the Learning Sciences, 24*(3), 347–372.

Battey, D., & Leyva, L. A. (2016). A framework for understanding whiteness in mathematics education. *Journal of Urban Mathematics Education, 9*(2), 49–80.

Bishop, R. S. (1990). Mirrors, windows, and sliding glass doors. *Perspectives: Choosing and Using Books for the Classroom, 1*(3), ix–xi.

Boaler, J. (1997). Setting, social class and survival of the quickest. *British Educational Research Journal, 23*(5), 575–595.

Boaler, J. (2002). *Experiencing school mathematics: Traditional and reform approaches to teaching and their impact on student learning.* Routledge.

Boaler, J., & Staples, M. (2008). Creating mathematical futures through an equitable teaching approach: The case of Railside School. *Teachers College Record, 110*(3), 608–645.

Brousseau, G. (2006). *Theory of didactical situations in mathematics: Didactique des mathématiques, 1970–1990* (Vol. 19). Springer Science & Business Media.

Bullock, E. C. (2018). Intersectional analysis in critical mathematics education research: A response to figure hiding. *Review of Research in Education, 42*(1), 122–145.

Calarco, J., Horn, I. S. & Chen, G. A. (2022). "You Need to be More Responsible": The Myth of Meritocracy and Teachers' Accounts of Homework Inequalities. Educational Researcher.

Chen, G. A. (2020). "That's obviously really insensitive:" Attuning to marginalization in a parent-teacher encounter. *Cognition and Instruction, 38*(2), 153–178.

Chen, G. A., & Buell, J. Y. (2018). Of models and myths: Asian (Americans) in STEM and the neoliberal racial project. *Race Ethnicity and Education, 21*(5), 607–625.

Coburn, C. E. (2001). Collective sensemaking about reading: How teachers mediate reading policy in their professional communities. *Educational Evaluation and Policy Analysis, 23*(2), 145–170.

Cohen, D. K. (2011). *Teaching and its predicaments.* Harvard University Press.

Cohen, E. G., & Lotan, R. A. (2014). *Designing groupwork: Strategies for the heterogeneous classroom* (3rd ed.). Teachers College Press.

Davidson, A. L. (1996). *Making and molding identity in schools: Student narratives on race, gender, and academic engagement.* SUNY Press.

de Araujo, Z., Roberts, S. A., Willey, C., & Zahner, W. (2018). English learners in K–12 mathematics education: A review of the literature. *Review of Educational Research, 88*(6), 879–919.

diSessa, A. A. (2006). A history of conceptual change research: Threads and fault lines. In R. K. Sawyer (Ed.), *The Cambridge Handbook of the Learning Sciences* (pp. 265–282). https://doi.org/10.1017/cbo9780511816833.017

Edwards, A. (2010). How can Vygotsky and his legacy help us to understand and develop teacher education. *Cultural-Historical Perspectives on Teacher Education and Development,* 63–77.

Ehrenfeld, N., & Horn, I. S. (2020). Initiation-entry-focus-exit and participation: A framework for understanding teacher groupwork monitoring routines. *Educational Studies in Mathematics, 103*(3), 251–272.

Eisenhart, M. (2021). The anthropology of learning revisited. *Anthropology & Education Quarterly, 52*(2), 209–221.

Emanuelsson, J., & Sahlström, F. (2008). The price of participation: Teacher control versus student participation in classroom interaction. *Scandinavian Journal of Educational Research, 52*(2), 205–223.

Erickson, F. (1982). Taught cognitive learning in its immediate environments: A neglected topic in the anthropology of education 1. *Anthropology & Education Quarterly, 13*(2), 149–180.

Esmonde, I., & Langer-Osuna, J. M. (2013). Power in numbers: Student participation in mathematical discussions in heterogeneous spaces. *Journal for Research in Mathematics Education, 44*(1), 288–315.

Fay, L. (2019, August 7). 74 Interview: Researcher Gloria Ladson-Billings on Culturally Relevant Teaching, the Role of Teachers in Trump's America, and Lessons From Her Two Decades in Education Research. Retrieved from https://www.the74million. org/article/74-interview-researcher-gloria-ladson-billings-on-culturally-relevant-teaching-the-role-of-teachers-in-trumps-america-lessons-from-her-two-decades-in-education-research/

Freire, P. (1970/2018). *Pedagogy of the oppressed.* Bloomsbury Academic.

Friedland, R., & Alford, R. R. (1991). Bringing society back in: Symbols, practices and institutional contradictions. In P. J. DiMaggio & W. W. Powell (Eds.), *The new institutionalism in organizational analysis* (pp. 232–266). University of Chicago Press.

Garner, B., Thorne, J. K., & Horn, I. S. (2017). Teachers interpreting data for instructional decisions: Where does equity come in? *Journal of Educational Administration, 55*(4), 407–426.

Gholson, M. L., & Martin, D. B. (2019). Blackgirl face: Racialized and gendered performativity in mathematical contexts. *ZDM, 51*(3), 391–404.

Giroux, H. A., & Penna, A. N. (1979). Social education in the classroom: The dynamics of the hidden curriculum. *Theory & Research in Social Education, 7*(1), 21–42.

Goffman, E. (1974). *Frame analysis: An essay on the organization of experience.* Harvard University Press.

Gresalfi, M. S. (2009). Taking up opportunities to learn: Constructing dispositions in mathematics classrooms. *The Journal of the Learning Sciences, 18*(3), 327–369.

Gutiérrez, K. D., & Rogoff, B. (2003). Cultural ways of learning: Individual traits or repertoires of practice. *Educational Researcher, 32*(5), 19–25.

Gutiérrez, R. (2002). Beyond essentialism: The complexity of language in teaching mathematics to Latina/o students. *American Educational Research Journal, 39*(4), 1047–1088.

Gutiérrez, R. (2016). Strategies for creative insubordination in mathematics teaching. *Special Issue Mathematics Education: Through the Lens of Social Justice.*

Gutiérrez, R. (2018). *Rehumanizing mathematics: A vision for the future.* Invited speaker at the Latinx in the Mathematics Sciences Conference.

Gutstein, E. (2006). *Reading and writing the world with mathematics: Toward a pedagogy for social justice.* Taylor & Francis.

Hall, R., & Horn, I. S. (2012). Talk and conceptual change at work: Adequate representation and epistemic stance in a comparative analysis of statistical consulting and teacher workgroups. *Mind, Culture, and Activity, 19*(3), 240–258.

Hall, R., & Jurow, A. S. (2015). Changing concepts in activity: Descriptive and design studies of consequential learning in conceptual practices. *Educational Psychologist, 50*(3), 173–189.

Hand, V. (2012). Seeing culture and power in mathematical learning: Toward a model of equitable instruction. *Educational Studies in Mathematics, 80*(1), 233–247.

Haraway, D. (1988). Situated knowledges: The science question in feminism and the privilege of partial perspective. *Feminist Studies, 14*(3), 575–599.

Henningsen, M., & Stein, M. K. (2002). Supporting students' high-level thinking, reasoning, and communication in mathematics. *Research, reflection, and practice,* 46–61.

Herbel-Eisenmann, B. A. (2007). From intended curriculum to written curriculum: Examining the voice of a mathematics textbook. *Journal for Research in Mathematics Education, 38*(4), 344–369.

Hintz, A., & Tyson, K. (2015). Complex listening: Supporting students to listen as mathematical sense-makers. *Mathematical Thinking and Learning, 17*(4), 296–326.

Horn, I. S. (2007). Fast kids, slow kids, lazy kids: Framing the mismatch problem in mathematics teachers' conversations. *The Journal of the Learning Sciences, 16*(1), 37–79.

Horn, I. S. (2010). Teaching replays, teaching rehearsals, and re-visions of practice: Learning from colleagues in a mathematics teacher community. *Teachers College Record, 112*(1), 225–259.

Horn, I. S. (2012). *Strength in numbers: Collaborative learning in secondary mathematics.* National Council of Teachers of Mathematics.

Horn, I. S. (2013). Teaching as problem solving. In Y. Li, & J. Moschkovich (Eds.), *Proficiency and beliefs in learning and teaching mathematics: Learning from Alan Schoenfeld and Günter Törner* (pp. 125–138). Springer.

Horn, I. S. (2017). *Motivated: Designing math classrooms where students want to join in.* Heinemann.

Horn, I. S. (2018). Accountability as a design for teacher learning: Sensemaking about mathematics and equity in the NCLB era. *Urban Education, 53*(3), 382–408.

Horn, I., & Gresalfi, M. (2021). Broadening participation in mathematical inquiry: A problem of instructional design. In C. Chinn, R. Duncan, & S. Goldman (Eds.), *International handbook of inquiry and learning.* Routledge.

Horn, I. S., & Kane, B. D. (2015). Opportunities for professional learning in mathematics teacher workgroup conversations: Relationships to instructional expertise. *Journal of the Learning Sciences, 24*(3), 373–418.

Horn, I. S., & Kane, B. D. (2019). What we mean when we talk about teaching: The limits of professional language and possibilities for professionalizing discourse in teachers' conversations. *Teachers College Record, 121*(6), 1–32.

Horn, I. S., & Little, J. W. (2010). Attending to problems of practice: Routines and resources for professional learning in teachers' workplace interactions. *American Educational Research Journal, 47*(1), 181–217.

Jackson, K., Garrison, A., Wilson, J., Gibbons, L., & Shahan, E. (2013). Exploring relationships between setting up complex tasks and opportunities to learn in concluding whole-class discussions in middle-grades mathematics instruction. *Journal for Research in Mathematics Education, 44*(4), 646–682.

Jansen, A., Cooper, B., Vascellaro, S., & Wandless, P. (2017). Rough-draft talk in mathematics classrooms. *Mathematics Teaching in the Middle School, 22*(5), 304–307.

Joseph, N. M. (Ed.). (2020). *Understanding the intersections of race, gender, and gifted education: An anthology by and about talented Black Girls and women in STEM.* IAP.

Joseph, N. M., Hailu, M. F., & Matthews, J. S. (2019). Normalizing Black girls' humanity in mathematics classrooms. *Harvard Educational Review, 89*(1), 132–155.

Jurow, S., Horn, I. S., & Philip, T. M. (2019). Re-mediating knowledge infrastructures: A site for innovation in teacher education. *Journal of Education for teaching, 45*(1), 82–96.

Kazemi, E., & Stipek, D. (2009). Promoting conceptual thinking in four upper-elementary mathematics classrooms. *Journal of Education, 189*(1–2), 123–137.

Korthagen, F. A. (2010). Situated learning theory and the pedagogy of teacher education: Towards an integrative view of teacher behavior and teacher learning. *Teaching and Teacher Education, 26*(1), 98–106.

Ladson-Billings, G. (2014). Culturally relevant pedagogy 2.0: aka the remix. *Harvard Educational Review, 84*(1), 74–84.

Lakoff, G. (2014). *The all new don't think of an elephant!: Know your values and frame the debate.* Chelsea Green Publishing.

Lambert, R. (2015). Constructing and resisting disability in mathematics classrooms: A case study exploring the impact of different pedagogies. *Educational Studies in Mathematics*, *89*(1), 1–18.

Lampert, M. (1990). When the problem is not the question and the solution is not the answer: Mathematical knowing and teaching. *American Educational Research Journal*, *27*(1), 29–63.

Lampert, M. (2001). *Teaching problems and the problems of teaching*. Yale University Press.

Langer-Osuna, J. M. (2011). How Brianna became bossy and Kofi came out smart: Understanding the trajectories of identity and engagement for two group leaders in a project-based mathematics classroom. *Canadian Journal of Science, Mathematics and Technology Education*, *11*(3), 207–225.

Lave, J. (1988). *Cognition in practice: Mind, mathematics and culture in everyday life*. Cambridge University Press.

Lefstein, A., & Snell, J. (2013). *Better than best practice: Developing teaching and learning through dialogue*. Routledge.

Lefstein, A., Snell, J., & Israeli, M. (2015). From moves to sequences: Expanding the unit of analysis in the study of classroom discourse. *British Educational Research Journal*, *41*(5), 866–885.

Levin, B. B., & Ammon, P. (1992). The development of beginning teachers' pedagogical thinking: A longitudinal analysis of four case studies. *Teacher Education Quarterly*, 19–37.

Lewis, E. O. (1908). The effect of practice on the perception of the Müller-Lyer illusion. *British Journal of Psychology*, *2*(3), 294.

Lewis, K. E. (2014). Difference not deficit: Reconceptualizing mathematical learning disabilities. *Journal for Research in Mathematics Education*, *45*(3), 351–396.

Leyva, L. A. (2017). Unpacking the male superiority myth and masculinization of mathematics at the intersections: A review of research on gender in mathematics education. *Journal for Research in Mathematics Education*, *48*(4), 397–433.

Leyva, L. A. & Alley, Z. D. (2021). "Speaking out more in class" and "talk[ing] less and less about my goals": A counter-storytelling of undergraduate Latina women's critical race-gendered epistemologies as mathematics students and aspiring engineers. In M. C. Shanahan, B. Kim, K. Koh, A. P. Preciado-Babb & M. A. Takeuchi (Eds.), *Learning sciences in conversation: Theories, methodologies, and boundary spaces*. Routledge.

Liljedahl, P. (2016). Building thinking classrooms: Conditions for problem-solving. In P. Felmer et al. (Eds.), *Posing and solving mathematical problems* (pp. 361–386). Springer.

Little, J. W. (2003). Inside teacher community: Representations of classroom practice. *Teachers College Record*, *105*(6), 913–945.

Louie, N. L. (2017). The culture of exclusion in mathematics education and its persistence in equity-oriented teaching. *Journal for Research in Mathematics Education*, *48*(5), 488–519.

McGee, E. O., & Martin, D. B. (2011). "You would not believe what I have to go through to prove my intellectual value!" Stereotype management among academically successful Black mathematics and engineering students. *American Educational Research Journal*, *48*(6), 1347–1389.

Mehan, H. (1979). *Learning lessons: Social organization in the classroom*. Cambridge, MA: Harvard University Press.

Mehta, J. D. (2013). The penetration of technocratic logic into the educational field: Rationalizing schooling from the progressives to the present. *Teachers College Record*.

Michaels, S., O'Connor, C., & Resnick, L. B. (2008). Deliberative discourse idealized and realized: Accountable talk in the classroom and in civic life. *Studies in Philosophy and Education, 27*(4), 283–297.

Moschkovich, J. (1999). Supporting the participation of English language learners in mathematical discussions. *For the Learning of Mathematics 19*(1), 11–19.

National Academies of Sciences, Engineering, and Medicine. (2020). *Changing expectations for the K-12 teacher workforce: Policies, preservice education, professional development, and the workplace.* National Academies Press.

Nasir, N., Cabana, C., Shreve, B., Woodbury, E., & Louie, N. (2014). *Mathematics for Equity: A Framework for Successful Practice.* Teachers College Press.

National Governors Association. (2010). *Common Core State Standards.* Washington, DC.

National Research Council, & Mathematics Learning Study Committee. (2001). *Adding it up: Helping children learn mathematics.* National Academies Press.

Pascoe, C.J. (2011). *Dude, you're a fag: Masculinity and sexuality in high school.* UC Press.

Philip, T. M. (2011). Moving beyond our progressive lenses: Recognizing and building on the strengths of teachers of color. *Journal of Teacher Education, 62*(4), 356–366.

Philip, T. M. (2019). Principled improvisation to support novice teacher learning. *Teachers College Record, 121*(4), 4.

Rittle-Johnson, B., Star, J. R., & Durkin, K. (2020). How can cognitive-science research help improve education? The case of comparing multiple strategies to improve mathematics learning and teaching. *Current Directions in Psychological Science, 29*(6), 599–609.

Rogoff, B., Paradise, R., Mejía Arauz, R., Correa-Chávez, M., & Angelillo, C. (2003). Firsthand learning through intent participation. *Annual Review of Psychology, 54,* 175–203.

Rosebery, A., Ogonowski, M., DiSchino, M., & Warren, B. (2010). "The coat traps all your body heat": Heterogeneity as fundamental to learning. *The Journal of the Learning Sciences, 19,* 322–357.

Saxe, G. B. (2012). Approaches to reduction in treatments of culture-cognition relations: Affordances and limitations. Commentary on Gauvain and Munroe. *Human Development, 55,* 233–242.

Sfard, A. (2019). Learning, discursive faultiness, and dialogic engagement. In N. Mercer, R. Wegerif & L. Major (Eds.), *The Routledge International Handbook on Dialogic Education,* pp. 88–99.

Shah, N. (2017). Race, ideology, and academic ability: A relational analysis of racial narratives in mathematics. *Teachers College Record, 119*(7), 1–42.

Smith, M. S., & Stein, M. K. (2011). *5 practices for orchestrating productive mathematics discussions.* National Council of Teachers of Mathematics.

Stein, M. K., Grover, B. W., & Henningsen, M. (1996). Building student capacity for mathematical thinking and reasoning: An analysis of mathematical tasks used in reform classrooms. *American Educational Research Journal, 33*(2), 455–488.

Stigler, J. W., & Hiebert, J. (2009). *The teaching gap: Best ideas from the world's teachers for improving education in the classroom.* Simon and Schuster.

Thornton, P. H., & Ocasio, W. (2008). Institutional logics. In R. Greenwood, C. Oliver, R. Suddaby, & K. Sahlin (Eds.), *The Sage Handbook of Organizational Institutionalism* (pp. 99–129). Sage Publishing.

Tyack, D., & Tobin, W. (1994). The "grammar" of schooling: Why has it been so hard to change? *American Educational Research Journal, 31*(3), 453–479.

Valenzuela, A. (1999). *Subtractive schooling: US-Mexican youth and the politics of caring*. SUNY Press.

Van den Heuvel-Panhuizen, M., & Drijvers, P. (2020). Realistic mathematics education. *Encyclopedia of Mathematics Education*, 713–717.

Walkerdine, V. (1998). *Counting girls out: Girls and mathematics* (Vol. 8). Psychology Press

Walkerdine, V. (1990). Difference, cognition, and mathematics education. *For the Learning of Mathematics*, *10*(3), 51–55.

Ward, C. J., Nolen, S. B., & Horn, I. S. (2011). Productive friction: How conflict in student teaching creates opportunities for learning at the boundary. *International Journal of Educational Research*, *50*(1), 14–20.

Warren, B., Vossoughi, S., Rosebery, A. S., Bang, M., & Taylor, E. V. (2020). Multiple ways of knowing: Re-imagining disciplinary learning. In *Handbook of the cultural foundations of learning* (pp. 277–294). Routledge.

Wolcott, H. F. (1982). The anthropology of learning. *Anthropology & Education Quarterly*, *13*(2), 83–108.

Yackel, E., & Cobb, P. (1996). Sociomathematical norms, argumentation, and autonomy in mathematics. *Journal for Research in Mathematics Education*, *27*(4), 458–477.

3

MECHANISMS FOR TEACHERS' CONCEPT DEVELOPMENT

The Premises for Teacher Learning (Chapter 1) capture teachers' subject-world relations in important ways, highlighting how teachers are positioned with respect to what they learn. In Chapter 2, we described a telos for teacher learning of ambitious and equitable instruction that involves shifts in three key conceptual domains. We noted that concepts do not solely reside within individuals; instead, conceptual resources are embedded in teachers' workplaces, with their own cultural logics that render certain practices common sense. We also described a consensus in prior research that teachers accomplished in responsive instruction exhibit integrated concepts about teaching. Integrated concepts support ecological reasoning that helps teachers consider interrelationships among different aspects of instruction. To complete our theory of teacher learning, we now turn to the last component of Lave's learning theory framework. In this chapter, we describe the mechanism for learning, answering the question of how learning takes place.

Before we dive in, we offer an important theoretical note about a complexity in developing a teacher learning theory: A telos, by its very nature, asserts normative views of instruction – a description of what learning *should* lead to. Such normative views challenge the Competing Visions of Quality Premise, which posits that there are multiple versions of "good teaching" circulating within and across schools. How do we develop a general theory of learning while maintaining a commitment to a certain kind of instruction? Does learning only happen when teachers' practice grows toward ambitious and equitable instruction? We think it would be a mistake to insist on this, as it would exclude a lot of learning, thereby misconstruing the phenomenon of interest.

DOI: 10.4324/9781003182214-4

Our way out of this theoretical tangle is to embrace the specificity of the telos described in the last chapter while aiming for a more generic mechanism in this chapter. Namely, to be phenomenologically accurate, our learning mechanism describes teachers' development of new instructional practices, even if they do not move toward our desired endpoint. Thus, learning telos and learning mechanism are analytically distinct, with the mechanism needing to account for multiple possible trajectories, even ones we would not endorse.

Teacher Learning as the Evolution of Pedagogical Judgment

Building on Vygotsky's ideas, we describe the mechanism for teacher learning as the evolution of pedagogical judgment (Horn, 2019), the interplay between pedagogical action, pedagogical reasoning, and pedagogical responsibility (Figure 3.1). As we have stated, teachers learn as they develop concepts about instructional practice. These concepts change when teachers integrate feedback that suggests that their current understandings and practices no longer support productive action (the Utility Premise). Discerning what constitutes productive action is thus central to facilitating teacher learning. Additionally, focusing on the evolution of pedagogical judgment helps us understand how teachers' conceptual agency grows over time. Given the tensions between conceptual embeddedness and conceptual agency (Chapter 1) and the counter-normativity of ambitious and equitable teaching (Chapter 2), understanding teacher development through this lens gives us greater purchase on understanding – and ultimately supporting – teacher learning.

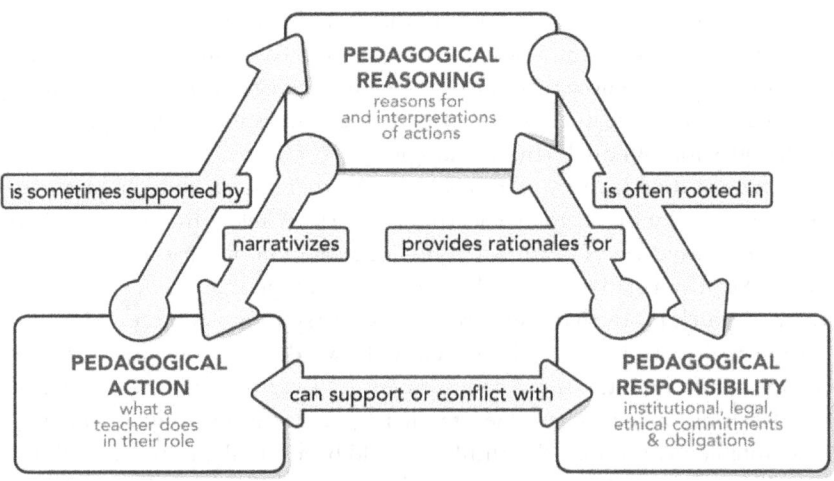

FIGURE 3.1 The components of pedagogical judgment and their relationships to one another.

Teachers' Discernment of Instructional Events

Miriam Sherin and Elizabeth van Es have richly documented teacher discernment by analyzing mathematics teachers' *noticing* (e.g., Sherin & van Es, 2009; van Es & Sherin, 2021). As they describe, teachers observe their classrooms, then select and make sense of aspects that they see as pedagogically relevant. As teachers become more adept at ambitious and equitable instruction, their discernment about what to pay attention to and what to disregard gets refined. They adopt inquiry stances and shift from teacher-centered views of classroom instruction to student-centered views.

In their framework, noticing and interpretation are crucial parts of teachers' work, constituting a *professional vision*. The idea of professional vision builds on sociolinguist Charles Goodwin's (1994) work examining archeologists and courtroom trials. He described how professionals interactionally code, highlight, and produce their material world, with experts helping novices develop specific ways of seeing ambiguous phenomena (e.g., in the case of archeologists, the precise classification for dirt samples).

While Sherin and colleagues have analyzed teacher noticing from a cognitive perspective, other scholars have extended their framework to encompass sociocultural and socio-material resources that shape teachers' noticing and professional vision (Nerland, 2012). For instance, Adam Lefstein and Julia Snell (2011) argue that professional vision has a political dimension, which benefits from analyses of power. This moves researchers away from framing discernment as normative questions of ability – teachers who are "more" or "less" able to notice the "right" things – toward a question of capacity shaped by social position, context, and commitments. As critical theorists remind us, normativity claims are deeply entangled with power claims (Butler, 1997; Foucault, 1975/2007; Harding, 2016).

Nicole Louie (2018) extended this situative perspective on noticing, showing how noticing is ideological as well as cognitive. Louie described how, as a teacher interpreted students' contributions in class discussions, she engaged ideologies that supported or inhibited noticing the value of students' ideas. Louie thus underscored the importance of the conceptual shift around who belongs in math class that we described in Chapter 2, since understanding multiple forms of mathematical competence is crucial for learning to value students' contributions. Further extending the situative perspective, Grace Chen (2020) examined pre-service teachers interacting with an actor playing a Kurdish refugee mother in a simulated parent-teacher conference, analyzing their capacities to listen to, respond to, or marginalize the mother as they discussed her child. Chen described the teachers' *attunements*, interactional sensitivities shaped – but not solely determined – by their own historicized experiences. Attunement highlights the affective dimensions of teacher discernment, the extent to which noticing and responding is an emotional – as well as ideological, social, and cognitive – process. Taken together, this literature underscores how teachers' conceptions, emotions, and lived experiences inform what is salient to them in the classroom, shaping their sense of what is and is not working in their instruction.

From Teacher Noticing to Pedagogical Judgment: Incorporating Salient Events into the Density of Decision-making

Teachers' interpretations of classroom events are a crucial component of the mechanism for teacher learning. As the Action and Interpretation Premises (Chapter 1) suggest, the organization of schooling requires that teachers act based on their interpretations of what is happening during instruction. As we and others have said, teachers make hundreds of judgments daily. Historian Larry Cuban's (1993) analysis of constancy and change in teaching illustrates the breadth of these judgments. They range from obvious judgments, like content and pedagogical strategies, to less obvious ones, like how students are allowed to move in the classroom. Teachers make decisions about assessing student knowledge, communicating with parents, regulating bathroom use, routines for homework, and so on. Teachers' pedagogical judgments go well beyond what can be captured in a concise, learnable template, especially since these decisions must make some sense within the logics governing particular teaching situations. Instruction is both too situationally bound and too broad to be adequately captured by lists of "best practices." Of course, some kinds of teacher learning can be accomplished by imitating more accomplished practitioners, and such approximations (Grossman & McDonald, 2008) can certainly scaffold deeper teacher learning as teachers reflect on their purposes and limitations. This dialectical relationship between action and reasoning is crucial in supporting the conceptual shifts described in Chapter 2, Teacher learning must go beyond incorporating new actions into developing new meaning systems – the narrativized actions and understandings that constitute concepts.

From a concept development perspective, we do not simply look for teacher learning as evidenced by the enactment of pre-determined practices; rather, we focus on teachers' pedagogical judgments, which link their practice to their understandings. Pedagogical judgments are highly consequential to students' experiences in school, an aspect of school life that is frequently overlooked by larger evaluation systems, yet matters for "functional validity" in measures of educational quality (Schneider et al., 2021). Teachers' decisions about allocating participation, for instance, play into dynamics of student inclusion, exclusion, and marginalization in the classroom (Alton-Lee et al., 1993; Spender, 1982), which, as we described in Chapter 2, shape students' long-term affiliation (or disaffiliation) with school (Anderman & Anderman, 1999; Goodenow, 1993). When biased, educators' judgments are a source of academic pushout (Morris, 2016), disproportionately excluding students from historically marginalized groups. As one example, discipline data in U.S. schools reveals that students of color, LGBTQ+ students, and students with disabilities are at greater risk for exclusionary discipline due to educators' biased interpretations of and responses to their behavior (Skiba et al., 2014). Messages about belonging are crucial for student engagement. They are especially important in mathematics, where student attrition is high (Boaler, 2000) – and even higher for women and non-binary students, first-generation college students, and students of color (Herzig, 2004; Noyes, 2009), all of whom are frequent targets of biased judgments.

We use a tripartite conception of pedagogical judgment, comprised of (1) pedagogical action supported by (2) pedagogical reasoning and rooted in (3) pedagogical responsibility. These three components are related but sufficiently distinct to warrant separate analysis. First, *pedagogical action* refers to the intentional and unintentional choices teachers make, both during and outside of classroom instruction. Examples of pedagogical action include giving tests, asking students to be quiet, conferring with colleagues, and emailing parents.

However, the same pedagogical action may have different interpretations and rationales – in our language, different underlying concepts – which become visible in teachers' *pedagogical reasoning* (Horn, 2005). Because the same action may arise for different reasons, pedagogical action and pedagogical reasoning are analytically distinct. For instance, teachers may give tests because they want to see what students know; they may be following departmental policies to test weekly; or their action may be rooted in both reasons.

Finally, *pedagogical responsibility* describes teachers' engagement with their sense of obligations, whether to ethical principles (Stengel & Casey, 2013; Tate, 2007) or situational constraints (Cuban, 1993; Herbel-Eisenmann et al., 2006). Thus, teachers may give tests because they feel responsible for preparing students for high-stakes assessments; they may differentiate the time allowed to respect special-education students' individualized educational plans; they may offer translations of word problems to students whose home languages differ from the language of instruction, reflecting a commitment to fair access. These choices respond to different types of pedagogical responsibilities – institutional, legal, and ethical.

Pedagogical responsibility is a highly personal and often underexamined aspect of pedagogical judgment. Teachers' pedagogical responsibility reflects their overarching commitments; these serve as both guides and anchors for pedagogical actions. Analytically, teachers' sense of pedagogical responsibility can be discerned through their visions of good teaching, which are a core part of teachers' *identities*. As teachers describe themselves and who they aim to be, these descriptions are shaped by things like their personal histories, social positions, and temperaments. Their self-descriptions use culturally available descriptors, narratives, and archetypes, which point to broader social constructions of teaching (Horn et al., 2008). For instance, teachers might describe themselves as "strict," pointing to commitments they have to their relationships with students and the kinds of classroom environments they value. Others might describe themselves as "liberatory," invoking another set of archetypes for relationships with students and classroom environments. In this way, teachers' identities provide descriptions of their individual dispositions and the social worlds they inhabit, reflecting their pedagogical responsibilities.

In describing what it means to know in teaching, Lee Shulman (1986) asserted that "[t]he teacher is not only a master of procedure but also of content and rationale, and capable of explaining why something is done" (p. 13). Yet we find that, even among thoughtful and reflective teachers, not all pedagogical actions are rooted in explicit pedagogical reasoning, let alone a clearly articulated sense of pedagogical responsibility. Instead, our studies of teachers' discourse reveal that teachers' reasons for their practices frequently remain tacit (Horn, 2005; Horn & Kane, 2015, 2019). For instance, teachers might give homework because it is a tradition of schooling: It is simply what teachers *do*, so there is no need to specify the underlying responsibility – giving homework is rooted in the institutional logics of schooling. At the same time, teachers can detail reasons for even mundane actions: They might assign homework to help students solidify their learning, or to communicate with parents what they are teaching.

Acknowledging this complexity, we generally find that actions that go against traditions of schooling demand rationales more frequently than actions that conform to institutional logics. For example, teachers who do not assign homework may need to articulate their pedagogical reasoning to justify their actions more frequently than teachers who do. This observation is supported by ethnomethodological theory (Garfinkel, 1967; Heritage, 2013), where it is commonly understood that practices that breach norms and expectations require explanation, while taken-for-granted practices do not. Because traditional instruction typically aligns with the institutional logics of schooling, counter-normative forms of instruction – which often includes instruction that is responsive, inclusive, and conceptually rich – places greater demands on teachers to explain themselves. This is likely why reflection has long been noted as an important dimension of teacher learning: Reflection helps teachers articulate reasons for non-conforming actions.

Generally, though, teachers' reasons may or may not connect to the broader institutional, ethical, or situational concerns that constitute pedagogical responsibility. Continuing the homework example, teachers may assign it for the pedagogical reason that children need independent practice. However, they may not have reflected on broader ethical issues of pedagogical responsibility, such as what homework might mean for students' after school obligations – or even their opportunities to freely play or pursue hobbies.

These examples illustrate that pedagogical action, pedagogical reasoning, and pedagogical responsibility constitute interrelated but independently analyzable components of pedagogical judgment.

Figure 3.1 summarizes this tripartite model of pedagogical judgment, detailing the relationships among the components. In the next section, we describe how they work together to support teachers' development of pedagogical judgment.

Illustrating the Development of Pedagogical Judgment

The mechanism we propose for teachers' development of pedagogical judgment considers the relationships among these three components. To illustrate with a hypothetical example, imagine a middle-grades math teacher, Mr. Green, who assigns nightly homework, rooted in a strong commitment that *homework is important for students' independent practice*. This onto-epistemic stance reflects his sense of pedagogical responsibility to help students master content. However, this year, he noticed that half of his students do not regularly turn in homework, which is creating problems for their grades. This datapoint is accompanied by an additional observation, which he incorporates into his interpretation: Students with sinking grades are becoming disengaged in class. He wonders if the pedagogical action of assigning homework is still useful (the Utility Premise), spurring potential paths for his learning.

One possible path for Mr. Green's learning is for him to revise his pedagogical reasoning without revising his pedagogical actions. In this scenario, Mr. Green's current narrative for his action − *homework is important for students' independent practice* − is no longer adequate, because he recognizes that his homework system also contributes to student disengagement. His newly formulated concept might be something like: *Homework is important for students' independent practice, but not all students are willing to complete it*. This new narrative illustrates a modest change, an interpretation that normalizes the situation without spurring him to revise his pedagogical action or responsibility. In fact, cultural mythology about mathematics supports this interpretation: Not everybody is a "math person," so it is expected that some students are unable to meet the challenge of mathematics class (Leyva et al., 2021).

A second possible path for Mr. Green's learning emphasizes revision to his pedagogical actions around homework, with small revisions to his pedagogical reasoning and responsibility. In this scenario, he might talk to a colleague, Ms. Honeydew, about her homework policies. She shares the webpage she uses to post her assignments, explaining the parent communication system she has built around it. Using these representations of practice, animated by replays of her explanations to students, Ms. Honeydew helps Mr. Green see how new pedagogical actions address his problem of practice. He revises his pedagogical reasoning with an additional diagnostic idea: *Students need additional support for homework completion through home-school communication.* His pedagogical responsibility – the commitment to students' independent practice – remains intact, with an additional commitment to improve communication with parents.

A third path involves revisions to Mr. Green's pedagogical actions, pedagogical reasoning, and pedagogical responsibility. Imagine he takes the homework problem to a different colleague, Ms. Ivory. Ms. Ivory explains that she thinks homework is not central to helping students learn, a direct contradiction to one of Mr. Green's core commitments about homework. Instead, her homework policy stems from a different vision of good teaching (Competing Visions of Quality), one rooted in her pedagogical responsibility to *create equitable access to learning.* Ms. Ivory explains that, several years ago, she noticed problems with her homework system, and this convinced her to change her approach. Specifically, she noticed clear social class differences between homework completers and non-completers. She realized that homework completers were mostly middle-class children whose parents had either sufficient education and time to help with homework or financial resources to afford tutors. This realization conflicted with her pedagogical responsibility to create equitable access to learning, and she saw that her old homework system contributed to inequalities in students' grades, leaving non-completers demoralized. Through several years of experimentation, she developed a weekly homework menu system, with a selection of practice problems that students self-assign. She explains how this system helps differentiate homework assignments for students – she sometimes suggests that students take on certain problems, but many problems are exploratory in nature so as to not create gaps among students who do different amounts of homework. She finds that her system ensures that important learning happens in class and de-emphasizes the connection between students' social class and grades.

Mr. Green's interaction with Ms. Ivory offers a different kind of learning opportunity. If he takes up her ideas, he would not only substantially shift his pedagogical action (de-emphasizing homework completion through self-guided homework menus), he would also have to reconsider a core principle of his pedagogical responsibility: the idea that students need homework for independent practice. In fact, it is very likely that integrating Ms. Ivory's ideas would shift Mr. Green's pedagogical reasoning about much of his teaching, since he will have to change how he assigns grades and thinks about developing students' computational fluency. This learning opportunity would press Mr. Green to reconcile aspects of his teacher identity and expand his repertoire of pedagogical actions – it would demand *negotiation* of who he is and who he wants to become as a teacher (Horn et al., 2008).

These imagined paths all involve Mr. Green contending with competing concepts about homework from his environment and making sense of them. In the first scenario, he notices his students' low homework completion rates and interprets this on his own, drawing on institutional norms (assigning homework is part of math class) and cultural myths (some people are not math people) to normalize the situation and modify his narrative about students and homework, signaling a new (if, in our view, problematic) conceptualization.

In the second two scenarios, Mr. Green's interactions with colleagues offer other feedback, interpretations, and possible actions for his problem of practice. These collegial interactions constitute learning opportunities (Horn et al., 2015) for Mr. Green, which he may or may not take up. Both Ms. Honeydew's and Ms. Ivory's proposals would require additional work on his part, which he may or may not determine to be worth the time investment. His subsequent actions will be shaped by his own sense of the usefulness of what they shared (Utility), as well as his personal vision of good teaching (Competing Visions of Quality). Teacher learning entails such negotiations, as teachers seek reasonable forms of practice amidst very real tensions – the non-ideal conditions that require teachers to respond in ways that are contextually relevant, morally adequate, and practically feasible (Jaggar, 2015).

Sometimes, teachers who are skeptical of colleagues' proposals overcome their hesitation if they respect or admire that person. In this way, the identity component of the Competing Visions of Quality guides and anchors teachers' practice and shapes their learning. In our hypothetical scenarios, Mr. Green's relationships with Ms. Honeydew and Ms. Ivory would influence his responses to these learning opportunities. If Mr. Green views one of these teachers as a role model, he is more likely to implement her suggested practice into his repertoire of pedagogical actions. Such identification is especially consequential when teachers are negotiating their core commitments, as Ms. Ivory's suggestion would require (Horn et al., 2008).

This example also highlights why dilemmas and conflicts are endemic to the development of pedagogical judgment. If Mr. Green considers Ms. Ivory's statement about homework exacerbating inequality, he would need to reconcile two competing aspects of pedagogical responsibility, one invested in students' independent practice and another invested in equitable learning environments. This conflict may spur new forms of pedagogical action, whether an adaptation of Ms. Ivory's homework menus or something altogether different. Such dilemmas offer another kind of concept development, one that potentially augments teachers' practice and presses them to reconcile competing pedagogical responsibilities (Chen et al., 2021; Ward et al., 2011).

Ultimately, experiences that support teachers' pedagogical reasoning about their pedagogical action in relation to their pedagogical responsibility offer a mechanism yielding new narratives of teaching events or practices. Pedagogical actions – what teachers do – do not always align with their pedagogical responsibility. Teachers have few opportunities compare their actions and responsibilities, especially in ways that support transformative integration of their conceptions of practice. Because of the Asynchronicity of Reflection, learning about teaching is difficult even when teachers *do* have time to work, reflect, and collaborate with colleagues. Often, discussions are hindered by weak representations of practice. Mr. Green's homework exploration was facilitated the material systems that Ms. Honeydew and Ms. Ivory could share. However, the interactive parts of teaching – like how Ms. Ivory addresses parents' and students' questions about her homework menus – are much harder to capture outside of the vivid action of instruction, which, in turn, makes them harder to reflect on and learn from (Hall & Horn, 2010).

Recall our description in Chapter 2 of the ways tasks often lose their richness in implementation: Teachers proceduralize the task in the set-up; they do not make clear connections to students' prior knowledge or experiences; they give inadequate time for students to explore; or one student dominates groupwork. Yet, in our studies of teachers' workgroup meetings (e.g., Horn et al., 2017), when teachers share why tasks "didn't work," they seldom consider the task set-up, pacing, or structures for student collaboration. To be sure, it is nearly impossible to conduct a lesson and notice all these details at the same time. As we argue more fully in Chapter 4, for reflection to support transformative teacher learning, it needs to be rooted in rich representations of practice that allow teachers to reason about the connections and disjunctures between their pedagogical actions and pedagogical responsibilities.

A Vygotskian Perspective on Teachers' Concept Development

It is difficult to develop a theory of teacher learning when the relationship between *learners* and *what they are learning* is both complex and socially embedded; it is even harder when images of good practice are highly contested. Amidst this messiness, we find that Vygotksy's theory helps us consider both the mechanism for teachers' learning and trajectories for development. Learning, for Vygotsky, entails learners' changing relationships with their social contexts. As learners grapple with what is socially valued, they develop new ways of interpreting and acting on the social world. Essentially, learning involves the *internalization* of *external* social processes. Vygotksy's emphasis on the social nature of knowledge and learners' agency in their development offers crucial resources for understanding teachers' learning. Specifically, we use his ideas about concept development to understand mechanisms for teacher learning.

To illustrate, consider the concept of *student engagement*. Teachers frequently consider issues of student engagement as they plan lessons, reflect on how activities went, or consider the strengths and challenges of different students. Student engagement is at the heart of any vision of teaching; without it, teachers' fundamental task of changing learners' understandings is impossible to accomplish. When teachers develop clear ideas of things like *student engagement* alongside causal explanations, a sense of the critical attributes, and a repertoire of related strategies, we say they have *developed a concept* about student engagement; practices are coupled with narratives of how they function. According to Vygotsky, meanings emerge from the dynamic interplay between these scientific concepts and lived concepts. Scientific concepts refer to formal, abstract ideas like *student engagement*, which might be linked to notions like "helping students become involved in lessons." Lived concepts refer to informal, everyday experiences related to those notions, such as images of students' body language or the pace and tone of class discussions. However, multiple meaning systems are at play in schools, and they may be more or less salient to teachers as they reflect on practice and decide how to refine their work – the sensitivity that Lefstein and Snell (2013) and Chen (2020) described. For some teachers, the lived version of *student engagement* might emphasize lively, active discussion; for others, it might center quiet students focused on tasks; it might even entail a combination, depending on the lesson goals. The clusters of these scientific and lived concepts, whether they are explicitly formulated or not, constitute teachers' understandings of different aspects of instruction, akin to Korthagen's (2010) gestalts for teaching. This relates to Vygotsky's emphasis on *perezhivanie* (1994) – lived or emotional experiences that are central to how people engage with and interpret the world.

As we stated earlier, we seek a learning mechanism that is non-normative, in that it explains different teacher learning trajectories, whether we consider them desirable or not. Our non-normative stance is rooted in commitments to viewing learning as an everyday phenomenon and from the humility of knowing that teachers constantly navigate non-ideal circumstances (Jaggar, 2015).

To this end, our studies use ethnomethodological approaches that center learners' sensemaking, as opposed to *a priori* notions of what we believe should constitute learning (Garfinkel, 1967; Heritage, 2013; Nasir et al., 2020). Undoubtedly, teachers' sensemaking sometimes runs counter to our goals as teacher educators but may meet other goals (like work-life balance or administrative priorities) that are also consequential. For instance, teachers may conclude that the slow and indeterminate work of *supporting student sensemaking* is not worth the tradeoff of falling behind district pacing guides. At other times, teachers' sensemaking helps us, as researchers and teacher educators, see ways to navigate the tensions between the goals of instructional practices and the institutional logics of their settings. The separation between our values as teacher learning theorists and our values as teacher educators creates space for expertise to flow both ways.

This non-normative stance is key to a sociocultural perspective on learning, as it helps analysts identify the role of contexts in meaning-making. The lack of consensus about what constitutes good teaching leaves ambiguity around what it means for teachers to "succeed," therefore making the broader values of schools and schooling an inevitable part of teachers' sensemaking.

Should Mr. Green take the institutionally sanctioned path of least resistance, and normalize low homework completion and student disengagement? Or should he listen to Ms. Honeydew or Ms. Ivory? In a longitudinal study of novice teachers' development, Susan Nolen and colleagues (2011) described how teachers' orientations toward good teaching changed as they adjusted to their first full-time teaching positions. In their study, one social studies teacher initially rejected the idea of developing rubrics, a practice encouraged in her teacher education program, because she felt they were "unrealistic" and "too much work." In her first teaching job, however, she found that her colleagues used and valued rubrics as tools, and, in collaboration with her new colleagues, she began creating and using rubrics to give student feedback. Her desire to identify with colleagues and be viewed as competent in her workplace helped her overcome her initial dislike of rubrics. In other words, as teachers learn ambitious and equitable instruction, over-ascribing their understandings to individual cognition misses important institutional, cultural, and relational contexts, which can explain aspects of their practice that do not conform to abstract, normative ideas. Vygotsky's theory, coupled with ethnomethodological commitments to non-normativity, invites us to see lived experiences as important parts of teachers' sensemaking.

Summary

This chapter outlined a general mechanism for teacher learning that draws on Vygotsky's notion of concept development as building connections between lived and scientific concepts. We explained the importance of distinguishing between telos and mechanism in our theory of teacher learning since the former invites normative narratives for teacher development and the latter cannot if it is empirically adequate (given the Competing Visions of Quality). We described how teachers develop pedagogical judgment as they work on problems of practice, issues that arise as they navigate non-ideal situations.

As teachers reason about these problems, lived concepts – which are often experienced as pedagogical actions – may be held up against scientific concepts – which are often used to narrate pedagogical reasoning and describe pedagogical responsibility. Depending on the adequacy of resources, teachers may productively diagnose and tinker with disjunctures they identify to develop new forms of pedagogical action, new reasonings, or new understandings of their pedagogical responsibilities – or a combination or integration of these. Pedagogical judgments are reorganized through expansion or integration, yielding new conceptions of key aspects of teaching.

One important limitation to teachers' learning comes from the Asynchronicity of Reflection, requiring teachers to anchor their reflections on inherently and inevitably incomplete versions of instruction. We take this up in our study design (Chapter 4).

References

Alton-Lee, A., Nuthall, G. & Patrick, J. (1993). Reframing classroom research: A lesson from the private world of children. *Harvard Educational Review, 63*(1), 50–85.

Anderman, L. H., & Anderman, E. M. (1999). Social predictors of changes in students' achievement goal orientations. *Contemporary Educational Psychology, 24*(1), 21–37.

Boaler, J. (2000). Mathematics from another world: Traditional communities and the alienation of learners. *The Journal of Mathematical Behavior, 18*(4), 379–397.

Butler, J. (1997). Merely cultural. *Social Text, 52/53*, 265–277.

Chen, G. A. (2020). "That's obviously really insensitive:" Attuning to marginalization in a parent-teacher encounter. *Cognition and Instruction, 38*(2), 153–178.

Chen, G. A., Marshall, S. A., & Horn, I. S. (2021). 'How do I choose?': mathematics teachers' sensemaking about pedagogical responsibility. *Pedagogy, Culture & Society, 29*(3), 379–396.

Cuban, L. (1993). *How teachers taught: Constancy and change in American classrooms, 1890–1990.* Teachers College Press.

Foucault, M. (1975/2007). *Discipline and punish: The birth of the prison.* Duke University Press.

Garfinkel, H. (1967) *Studies in ethnomethodology.* Paradigm Publishers.

Goodenow, C. (1993). The psychological sense of school membership among adolescents: Scale development and educational correlates. *Psychology in the Schools, 30*(1), 79–90.

Goodwin, C. (1994). Professional vision. *American Anthropologist, 96*, 606–633.

Grossman, P., & McDonald, M. (2008). Back to the future: Directions for research in teaching and teacher education. American Educational Research Journal, 45(1), 185–205.

Hall, R., & Horn, I. S. (2012). Talk and conceptual change at work: Adequate representation and epistemic stance in a comparative analysis of statistical consulting and teacher workgroups. *Mind, Culture, and Activity, 19*(3), 240–258.

Harding, S. (2016). *Whose science? Whose knowledge?* Cornell University Press.

Herbel-Eisenmann, B. A., Lubienski, S. T., & Id-Deen, L. (2006). Reconsidering the study of mathematics instructional practices: The importance of curricular context in understanding local and global teacher change. *Journal of Mathematics Teacher Education, 9*(4), 313–345.

Heritage, J. (2013). *Garfinkel and ethnomethodology.* John Wiley & Sons.

Herzig, A. H. (2004). Becoming mathematicians: Women and students of color choosing and leaving doctoral mathematics. *Review of Educational Research, 74*(2), 171–214.

Horn, I. S. (2005). Learning on the job: A situated account of teacher learning in high school mathematics departments. *Cognition and Instruction, 23*(2), 207–236.

Horn, I. S. (2019). Supporting the development of pedagogical judgment. *International Handbook of Mathematics Teacher Education: Volume 3: Participants in Mathematics Teacher Education, 321.*

Horn, I. S., Garner, B., Kane, B. D., & Brasel, J. (2017). A taxonomy of instructional learning opportunities in teachers' workgroup conversations. *Journal of Teacher Education, 68*(1), 41–54.

Horn, I. S., & Kane, B. D. (2015). Opportunities for professional learning in mathematics teacher workgroup conversations: Relationships to instructional expertise. *Journal of the Learning Sciences, 24*(3), 373–418.

Horn, I. S., Kane, B. D., & Wilson, J. (2015). Making sense of student performance data: Data use logics and mathematics teachers' learning opportunities. American Educational Research Journal, 52(2), 208–242.

Horn, I. S., & Kane, B. D. (2019). What we mean when we talk about teaching: The limits of professional language and possibilities for professionalizing discourse in teachers' conversations. *Teachers College Record, 121*(4), 4.

Horn, I. S., Nolen, S. B., Ward, C., & Campbell, S. S. (2008). Developing practices in multiple worlds: The role of identity in learning to teach. *Teacher Education Quarterly, 35*(3), 61–72.

Jaggar, A. M. (2015). Ideal and nonideal reasoning in educational theory. *Educational Theory, 65*(2), 111–126.

Korthagen, F. A. (2010). Situated learning theory and the pedagogy of teacher education: Towards an integrative view of teacher behavior and teacher learning. *Teaching and Teacher Education, 26*(1), 98–106.

Lefstein, A., & Snell, J. (2011). Professional vision and the politics of teacher learning. *Teaching and Teacher Education, 27*(3), 505–514.

Lefstein, A., & Snell, J. (2013). *Better than best practice: Developing teaching and learning through dialogue*. Routledge.

Leyva, L. A., McNeill, R. T., Marshall, B. L., & Guzmán, O. A. (2021). "It seems like they purposefully try to make as many kids drop": An analysis of logics and mechanisms of racial-gendered inequality in introductory mathematics instruction. *The Journal of Higher Education*, 1–31.

Louie, N. L. (2018). Culture and ideology in mathematics teacher noticing. *Educational Studies in Mathematics, 97*(1), 55–69.

Morris, M. (2016). *Pushout: The criminalization of Black girls in schools*. The New Press.

Nasir, N., McKinney de Royston, M., Barron, B., Bell, P., Pea, R., Stevens, R. & Goldman, S. (2020). Learning Pathways: How learning is culturally organized. In N. S. Nasir, C. D. Lee, R. Pea, and M. McKinney de Royston (Eds.), *Handbook of the cultural foundations of learning* (pp. 195–211). Routledge.

Nerland, M. (2012). Professions as knowledge cultures. In *Professional learning in the knowledge society* (pp. 27–48). Brill Sense.

Nolen, S. B., Horn, I. S., Ward, C. J., & Childers, S. A. (2011). Novice teacher learning and motivation across contexts: Assessment tools as boundary objects. *Cognition and Instruction, 29*(1), 88–122.

Noyes, A. (2009). Exploring social patterns of participation in university-entrance level mathematics in England. *Research in Mathematics Education, 11*(2), 167–183.

Schneider, J., Noonan, J., White, R. S., Gagnon, D., & Carey, A. (2021). Adding "Student Voice" to the Mix: Perception Surveys and State Accountability Systems. *AERA Open*. https://doi.org/10.1177/2332858421990729.

Sherin, M., & Van Es, E. A. (2009). Effects of video club participation on teachers' professional vision. *Journal of teacher education, 60*(1), 20–37.

Shulman, L. S. (1986). Those who understand: Knowledge growth in teaching. *Educational Researcher, 15*(2), 4–14.

Skiba, R. J., Arredondo, M. I., & Rausch, M. K. (2014). *New and developing research on disparities in discipline*. Bloomington, IN: The Equity Project at Indiana University.

Spender, D. (1982). *Invisible women: The schooling scandal*. Writers and Readers Publishing Cooperative Society.

Stengel, B. S., & Casey, M. E. (2013). "Grow by looking": From moral perception to pedagogical responsibility. *Yearbook of the National Society for the Study of Education, 112*(1), 116–135.

Tate, P. M. (2007). Academic and relational responsibilities of teaching. *The Journal of Education, 187*(3), pp. 1–20.

van Es, E. A., & Sherin, M. G. (2021). Expanding on prior conceptualizations of teacher noticing. ZDM–Mathematics Education, 53(1), 17–27.

Vygotsky, L. (1994). The problem of environment. In R. van der Veer, & J. Valsiner (Eds.), *The Vygotsky reader*, pp. 338–354. Blackwell.

Ward, C. J., Nolen, S. B., & Horn, I. S. (2011). Productive friction: How conflict in student teaching creates opportunities for learning at the boundary. *International Journal of Educational Research, 50*(1), 14–20.

PART 2
Studying Mathematics Teachers' Learning

In Part 1, we argued that learning to teach mathematics equitably and ambitiously is a conceptual and cultural change project. We synthesized prior research on the conditions and sociocultural processes of teacher learning, offering our theory of teachers' concept development by outlining their subject-world relations (how teachers are positioned in relation to what they know), telos (how concepts shift to support ambitious and equitable instruction), and mechanism (how those shifts happen).

In Part 2, we transition to our empirical investigation using this theoretical framework. Our central research question was: *How can we use formative feedback to enhance teachers' learning of ambitious and equitable mathematics instruction in urban secondary schools?* We offer some answers in this part of the book.

In Chapter 4, using our perspective on teacher learning, we discuss current research and practice of professional development. Building on our critiques of this literature, we introduce the design for Project SIGMa, explaining how we pursued our question through a research-practice partnership with a "best case" sample of teachers. We also describe our intervention – the video formative feedback (VFF) cycle – and the design conjectures that shaped it. Finally, we describe the teachers who participated in our two-year intervention and the data we collected to document their learning.

In Chapter 5, we explain what SIGMa teachers experienced in the VFFs and how we facilitated them. To illustrate how VFFs supported teachers' reconceptualization of their practice, we show how teachers' inquiry unfolded through one debrief conversation. Stepping back, we share our summative coding of the VFF debriefs, which captures the issues in ambitious and equitable instruction that this intervention helped teachers investigate.

DOI: 10.4324/9781003182214-5

As other researchers have noted, inquiry into teaching is uncommon for most teachers. Teachers tend to use evaluative discourse when viewing their own or others' lessons. However, aligned with our situative concept development perspective, we wanted to foster exploratory discourse to engage with teachers' pedagogical reasoning.

We thus devote Chapter 6 to the different ways teachers learned to learn in and through VFFs. We argue that shifting from evaluation to interpretation required that teachers let go of binary ideas about teaching's "goodness." Instead of viewing teaching practices as "good" or "bad," we encouraged a more expansive perspective, asking what and whom certain practices might be good *for*. This shift from global judgments to particular consequences renders teaching more discussable, and, in turn, it surfaced teachers' (sometimes conflicting) pedagogical responsibilities. We illustrate how this unfolded for three SIGMa teachers, arguing that the expansive stance on teaching supported their concept development.

In Chapters 7 and 8, we delve into specific cases of teachers learning about ambitious and equitable mathematics instruction. In Chapter 7, we focus on how VFFs produced moments of insight. Using rich representations of practice around teachers' inquiry questions, the SIGMa team facilitated conversations that allowed teachers to juxtapose their pedagogical actions and pedagogical responsibilities. When these were not aligned, we helped teachers reconceptualize actions that would better fit their commitments and goals, often through diagnostic conversations rooted in the video and through brainstorms about alternative ways of working in the future. We show three cases of teachers' learning on this relatively brief timeline.

Although many SIGMa teachers experienced brief yet meaningful episodes of learning during VFFs, they also sustained inquiry into practice over time. In Chapter 8, we share two cases of teacher learning on broader timescales, with particular attention to how the VFFs enabled that learning. By illustrating the iterative process of concept development, these cases show how teachers integrated their understandings about important aspects of ambitious and equitable mathematics instruction.

4

DESIGNING TO SUPPORT MATHEMATICS TEACHERS' CONCEPTUAL CHANGE

What We Know About Professional Development and Teacher Learning

In this chapter, we use our framework for teacher learning to critique current research on professional development and ground our study's design. To develop robust concepts about ambitious and equitable instruction, teachers need support. In the U.S., the professional development "system" – ad hoc workshops, institutes, and study groups – does not usually offer resources for this kind of learning. Most U.S. teachers report that their primary professional development experiences consist of brief, episodic workshops that are disconnected from practice (Wei et al., 2010). A recent survey found that about 40% of secondary mathematics teachers reported participating in at least 36 hours of mathematics-focused professional development in the past three years, with those teaching historically marginalized students more likely to participate in professional development (Banilower et al., 2018). At the same time, we know little about the quality of teachers' professional development, let alone if it supports ambitious and equitable mathematics instruction. Importantly, only about 20% of mathematics teachers reported attending professional development on incorporating students' cultural backgrounds into their teaching (Banilower et al., 2018), a critical issue for equity. While the current professional development environment emphasizes mathematics content, it does not address the importance of responding to students' backgrounds, suggesting that it is inadequate for supporting teacher learning about ambitious and equitable mathematics instruction.

Moreover, large-scale research has repeatedly shown that typical professional development interventions have limited influence on teachers' instructional practices (Garet et al., 2001; Porter et al., 2003). Professional development research has been critiqued for relying on self-reports and paying insufficient attention to the quality of teachers' learning experiences. The few

DOI: 10.4324/9781003182214-6

studies that use observational methods report that instructional change is often limited to surface features of practice (Grigg et al., 2013). Even fewer studies link professional development experiences to student learning (Kennedy, 2016; Lynch et al., 2019), a crucial outcome.

Despite these shortcomings, the field has developed consensus around what constitutes effective professional development. Professional development designs that have significant, positive effects on instruction share identifiable characteristics. In all, prior work shows that high-quality professional development:

- Focuses on teachers' subject-matter knowledge (Garet et al., 2001; Porter et al., 2003);
- Is organized around materials teachers use in their classrooms (K. Jackson & Cobb, 2013);
- Focuses on specific, effective instructional practices (Desimone et al., 2002);
- Creates opportunities for teachers' active learning (Garet et al., 2001; Porter et al., 2003);
- Is coherent with teachers' other learning activities (Garet et al., 2001; K. Jackson et al., 2018; Porter et al., 2003);
- Garners support from teacher communities (K. Jackson & Cobb, 2013; Wilson & Berne, 1999);
- Is sustained over time (Darling-Hammond et al., 2017; K. Jackson & Cobb, 2013).

Echoing a key argument in Mary Kennedy's (2016) metasynthesis of professional development and student learning, we note that, while checklists like this are good starting points for designs, they do not tell us how components might work synergistically to support teacher learning. This is especially true when designers aim to support teachers' concept development of ambitious and equitable instruction, which requires more than superficial changes in practice. What is the balance of, say, focusing on content knowledge versus specific instructional practices in a workshop? Although content knowledge is often referenced as an important component of professional development (Desimone, 2009), Kennedy found that, across disciplines, programs that focused exclusively on content generally had little effect on student learning. Somewhat contradictorily, Kathleen Lynch and colleagues' (2019) meta-synthesis of STEM instructional improvement found that an emphasis on content knowledge and pedagogical content knowledge supported better outcomes for students. These subtly different findings could be due to the importance of discussing content knowledge *alongside* its pedagogical use or due to STEM teachers' greater need for continued content learning (the subject of Lynch and colleagues' synthesis) compared to teachers in general.

Other findings from Kennedy's analysis were more counterintuitive when held against the folklore of professional development. For instance, *duration* and *intensiveness* of professional development interventions are taken-for-granted as positive design features, yet Kennedy found that program intensity alone (which has multiple meanings, including contact hours, span of time over which hours are distributed, and volume of information) did not yield better student learning outcomes. This was especially true for prescriptive professional development designs – ones that reduce teacher discretion and instead present instructional strategies as universally applicable. This last finding supports our commitment to teachers' concept development: Prescriptive designs, by definition, seek teachers' compliance with practices rather than cultivating their agency to support their interpretations and sensemaking.

Professional development research also points to the need for teacher learning activities to provide active, ongoing opportunities for inquiry that are close to everyday instruction and coordinated with other parts of teachers' work. To that end, instructional coaching has become an increasingly popular form of school-based professional development (Gallucci et al., 2010). However, coaching is often a limited resource; some studies have shown that full-time coaches only spend between one-quarter and one-third of their time working with teachers (Bean et al., 2010; Kane & Rosenquist, 2019; Ponglinco et al., 2003). Our informal observations in schools suggest that instructional coaches primarily work with early-career teachers and teachers who need extra support rather than their more experienced colleagues; in other words, instructional coaches tend to raise the floor, not the ceiling, of instructional practice. Furthermore, instructional coaching can be socially and emotionally risky (Schneeberger McGugan et al., 2021; Zembylas, 2003). When teachers allow coaches into their classrooms – even for non-evaluative observations – they are opening their instructional practice for critique. Instructional coaching can support conceptual change, particularly as teachers integrate scientific and lived concepts to develop more nuanced understandings of teaching. But at the same time, the vulnerability involved in observation and feedback cycles can make instructional coaching an uncomfortable experience.

Overall, research on professional development offers important insights about possibilities for teacher learning. From our conceptual change perspective, other pressing design questions arise. For example, we can ask: *How do these individual features of professional development address teachers' extant ideas about what they are learning? How might professional development transform teachers' conceptions about teaching?* As our brief review has illustrated, professional development research offers evidence for including some potentially useful features, but these lists do not constitute adequate frameworks for designing for professional learning because they lack an underlying theory – a longstanding issue in teacher education (Ball & Cohen, 1990; Goldsmith et al., 2014; Kennedy, 2016; Lynch et al., 2019). This is a serious limitation. Just as having better theories of student learning supports strategic instructional approaches, so too would better theories of teacher learning inform strategic professional development. Stronger theories about the nature of teacher learning could help us translate lists of characteristics into specific approaches and better specify the conditions under which teacher learning happens. Additionally, good theory lets us accumulate knowledge in the field, supporting empirically-grounded refinements in professional development designs and, ultimately, stronger influences on instructional practice.

Designing for Mathematics Teacher Learning

Identifying a Site to Design for and Study Teacher Learning

To extend current understandings of teacher learning, the Project SIGMa team used several strategies to find a suitable site to research these issues. First, we sought out a place where our phenomenon of interest – secondary mathematics teachers' learning about ambitious and equitable instruction – would be visible. Second, we wanted to collaborate with a program that already reflected features of effective professional development and had evidence of using them well; this would enable us to extend the field's knowledge. Third, to follow teachers' learning over time, we hoped to work with teachers who were likely to remain in their schools, since participant turnover is common in studies involving teachers – especially in U.S. schools serving historically marginalized students, which often have unfavorable working conditions that contribute to teacher mobility (Ingersoll & May, 2012). Finally, we wanted our design to be sustainable by building it into an organization, responding to the common critique that innovations for teacher development are seldom maintained once external funding and research personnel go away.

Finding a Place Replete with Teachers Learning About Ambitious and Equitable Math Instruction

To meet our first criterion, we partnered with a professional development organization (PDO) in a large urban school district. The PDO leaders' vision for mathematics teaching strongly aligned with ambitious and equitable instruction (Chapter 2). Teachers applied to the PDO for multi-year fellowships in school-based teams; in their applications, teams identified particular goals for their collective improvement. The PDO then provided each teacher with a stipend; money for their team's improvement project; a collaborative planning period that reduced their teaching load and gave them time to collaborate during their workweeks; support for travel to professional conferences; and monthly professional development workshops with the other fellows, who totaled about 100 in all. At the monthly full-day workshops, teachers participated in various activities. They attended short (30- to 60-minute) sessions on topics like *Group Assessments, Making Tasks Richer, Promoting Understanding in Troubling Times,* and *School Mathematics for Social Change.* They also joined year-long working groups, which were often organized by grade level or content areas (e.g., *7th Grade Math, Geometry,* and *AP Calculus*), with some dedicated to cross-cutting problems of practice (e.g., *Assessment, Motivation,* and *Classroom Discourse*). Working groups met monthly, allowing teachers repeated opportunities to collaborate with colleagues from other schools. Undoubtedly, the PDO's *raison d'être* was mathematics teacher learning, with a commitment to supporting teachers' work with their diverse students. Over 90% of the district's students came from groups that have been historically racially, linguistically, or culturally minoritized in the U.S.; approximately 80% of students qualified for free or reduced-price meals, indicating a high number of low-income families. Because of the students they served and the PDO leaders' commitments, we knew that issues of equity and inclusion would be central to conversations about ambitious mathematics instruction.

Reflecting What is Known About High-Quality Professional Development

The PDO's programming met our second criterion, reflecting the features of effective professional development described in the research literature. The monthly workshops were divided into several sessions. Reflecting the features of effective professional development, one session always *focused on content knowledge,* offering deep-dive problem-solving sessions on topics like *Graph Theory, the Mysteries of the Equilateral Triangle,* or *Markov Chains.* These sessions were led by the PDO's mathematician and were consistent teacher favorites, the room filling with a pleasant hum of engaged activity as teachers worked together on problem sets and explored mathematical ideas. Teachers also attended a three-week residential summer program that, in addition to its focus on mathematics instruction, allowed them sustained, inquiry-based learning experiences on topics like *Probability and Big Data* or *Applications of Geometric Thinking.* In interviews, teachers often mentioned that these mathematical experiences offered them insight into what their students experienced as mathematical learners, reminding them of how it feels to explore novel ideas and how pedagogical designs can facilitate learning.

The monthly workshops were also *organized around materials used in the classroom* and *focused on specific instructional practices.* In some sessions, PDO teachers presented on their classroom practices or things they learned at conferences, offering examples of their work on topics like integrating technology, involving reluctant students, and giving meaningful feedback. The monthly workshops and summer program, as well as their work on school-based improvement projects, demanded *active learning* from teachers. Because the teacher fellows were selected with a stated commitment to ambitious and equitable instruc-tion, there was greater-than-typical *coherence* between the work they did in their schools and the aims of the PDO activities. Furthermore, PDO affilia-tion *garnered support from teacher communities,* both at the school level through the school-based teams, as well as more broadly, by supporting collegiality across the fellows. Finally, because teachers applied for the PDO, this signaled their motivation to learn, a condition which undoubtedly enhanced their uptake of the PDO's promoted instructional approaches (Kennedy, 2016). In other words, the PDO participants already expressed and demonstrated com-mitments to ambitious and equitable mathematics instruction, meaning that their sense of pedagogical responsibility oriented them toward this form of teaching, increasing the likelihood that we, as researchers, could uncover and document this learning.

Reducing Participant Turnover

Regarding the third criterion, the PDO's fellowship structure stabilized our participant pool. Because the PDO fellowships lasted for five years and were contingent upon teaching in a "high-need" school, studying teachers within the PDO reduced participant turnover, aiding in the development of longi-tudinal cases of teachers' learning. Indeed, although two of our participants changed schools, they stayed within our focal teacher teams, making for negli-gible participant attrition. In comparison, in our last study based in U.S. urban schools, we tried to build multi-year cases of teacher communities but found ourselves having to conduct site selection every year due to teacher attrition and organizational churn (Horn et al., 2017).

Sustaining Designs Within the Organization

In terms of our fourth criterion, as we explored the possibility of this project, it was quickly evident that our research team and the PDO leaders shared a mutual commitment to supporting teachers' learning of ambitious and equi-table mathematics instruction, with a shared focus on designing professional learning activities to support that work. Our partners were eager to share and learn alongside us, exhibiting a strong investment in ongoing improvement.

To ground our partnership work in a shared question, we reviewed reports from the PDO's external evaluators. We learned that, overall, PDO teachers were pleased with the support and resources that they received; however, they wished for more direct coaching and mentoring tailored to their instruction. As we conceptualized and refined our intervention, we aimed to under-stand it well enough to offer a modified version to our PDO partners so that they could continue using it to meet their goals. In the meantime, we also served as "eyes and ears on the ground" for the PDO leaders, giving them de-identified feedback of what we saw in teachers' responses to and uptake of the various activities they offered, thereby acting as partners on other aspects of their professional development work. For instance, as we observed across multiple monthly workshops, we noted that the quality of teacher discussion varied in the yearlong working groups. We worked with our PDO partners to brainstorm structures that might ensure more consistently high-quality conversations across the groups. Finally, members of our research team shared our expertise, offering sessions at the monthly workshops to both support our partners and become more central participants in the organization. In the end, the PDO met our hopes as a research site. It was indeed a fruitful place for our investigation of secondary mathematics teachers' learning.

How Teachers Learn: Conceptualizing Teachers' Professional Learning Environments

The teacher learning theory we outlined in Part 1 offers conceptions of (1) subject-world relations; (2) telos for development; and (3) mechanism for learning. Applying these ideas to a professional development design required us to consider what a productive learning environment might look like in light of our theory.

Consonant with our sociocultural perspective, we consider the roles of both formal professional development activities and schools-as-workplaces to conceptualize teachers' learning environments (Kazemi & Hubbard, 2008). Introducing the school-as-workplace – which we captured in the learning conditions for teachers (Chapter 1) – extends the scope of typical professional development designs, which tend to focus solely on activities organized by professional development providers. Taking a broader view is key to our sociocultural perspective; it echoes similar analytic moves in contemporary research on student learning. Consider, for example, Alan Schoenfeld's (1988) findings about the influence of the hidden curriculum of schooling (P. Jackson, 1990) on students' mathematical learning. Specifically, Schoenfeld's analysis of the mathematical disasters of "good teaching" illustrated how students develop unproductive beliefs and practices by doing math that is infused by this hidden curriculum, such as believing that all math problems should be solvable in two minutes or less. Similarly, a sociocultural perspective insists that teachers' workplace norms constitute a hidden curriculum for their learning, social processes which reflect the institutional nature of schooling which, when reinstated alongside their "peculiar" understandings, uncover the rational basis of their sensemaking. For instance, many elementary teachers regularly assign their students math homework because it is a tradition of schooling and many parent communities expect this as a sign of rigor; however, research shows repeatedly that elementary math homework does not aid student learning and, in fact, mostly exacerbates class- and race-based inequities (Muhlenbruck et al., 2000).

To incorporate teachers' informal learning in schools, we build on our earlier research (Horn, 2005; Hall & Horn, 2012; Horn & Little, 2010; Horn & Kane, 2015; Horn et al., 2020). As we described in Chapter 2, this work identified how schools constitute important learning environments for teachers. Ordinary aspects of school life – curricular artifacts, reform slogans, descriptions like "fast" and "slow" kids – provide teachers with interpretive resources that constitute a conceptual infrastructure. Recall that we described how this conceptual infrastructure – institutional organization, cultural practices, and representations – shapes activities and distributes concepts across people and things, guiding teachers' actions and interpretations. This situative perspective on teachers' cognition and action not only pushes back on dominant theories of teacher learning and teacher change, it also counters broader cultural mythologies: Despite policy rhetoric (e.g., Kane & Staiger, 2008) and Hollywood movies (Breault, 2009) that emphasize individual teachers as the sole locus of instructional efficacy and agency, research on teachers' workplace learning uncovers how instructional practices are negotiated with and constrained by institutional norms.

Taking this situative perspective on teachers' learning, we needed to conceptualize the PDO as a learning environment and then develop some ideas about how to enhance it. In this approach – known as design-based research (Sandoval & Bell, 2004) – researchers develop principled conjectures about how particular designs support learning, going beyond a checklist of desired features toward principles that allow us to examine our notions of teachers' learning.

To ground our conjectures, we adapted John Bransford, Ann Brown, and Rodney Cocking's (1999) *How People Learn* framework. We first used this framework to examine how the PDO activities operated as a learning environment to support teachers' development of ambitious and equitable instruction. Then, we did a gap analysis, noting any places for potential improvement, a process called *conjecture mapping* (Sandoval, 2014). After our pilot year, we added an additional conjecture to reflect our emerging practice. We used these design conjectures to develop our intervention. By specifying the conjectures, we could systematically investigate how the learning environment influenced teachers' learning trajectories and then refine our understanding of their learning.

Centering Teachers as Learners

Bransford and colleagues explain that, whether we are talking about students or teachers, learning environments' effectiveness depends on how they center *learners, knowledge, assessments,* and *communities*. First, to center learners, environments attend carefully to the prior knowledge that learners bring to educational experiences. Our perspective considers teachers' prior knowledge to encompass their personal histories, cultural identities, experiences, and institutional contexts. Understanding learners' starting place is critical to concept development. For example, we know that teachers' own experiences in school strongly shape their images and expectations for their own work (Lortie, 1976/2020). This is even more true for experienced teachers, since they have developed habits and ideas about what teaching *should* and *can* look like. Learner-centered environments assume that learners construct their own meanings, so learning activities need to engage with and build on these. Yet in most professional development, this rarely happens. As we noted, traditional professional development seldom attends to teachers' own meanings, leading to conceptual slippage (Horn & Kane, 2019), where new ideas and practices are simply assimilated into old instructional models, leaving teaching fundamentally unchanged (Cohen, 1990; Schneider, 2014). Furthermore, learner-centered environments engage learners' goals for their learning.[1] Although some districts and schools are experimenting with differentiated professional development where teachers select from a menu of offerings (Sackstein, 2015), it is unclear how such programs would attend to knowledge construction, rather than merely accommodate personal preference. In sum, learner-centered professional learning environments:

- Address teachers' existing concepts about and practices for teaching (*Design Conjecture 1*)
- Align with teachers' personal goals for their learning (*Design Conjecture 2*)

Centering Teacher Knowledge

Second, environments that center knowledge attend to notions of competence – in this case, what does accomplished teaching look like? Mathematics education research is replete with examples that provide images of ambitious and equitable teaching, particularly in K-8 classrooms. For instance, Deborah Ball and Magdalene Lampert richly documented their 3rd- and 5th-grade teaching, building archives of classroom practice and student work (Lampert & Ball, 1998). Video cases are available on the Internet (e.g., the Teaching Channel, YouCubed.com) and in books (e.g., Boaler & Humphreys, 2005; Munson, 2018). Beyond images of classrooms, practitioner books and articles detail approaches for addressing challenging aspects of ambitious and equitable instruction, such as leading discussions (Smith & Stein, 2011), supporting student collaboration (Horn, 2012), fostering motivation (Horn, 2017), and using rich tasks (Henningsen & Stein, 1997). These resources offer visions of ambitious and equitable mathematics instruction, yet they do not always overcome teachers' sense that certain practices only work in *some* settings. In a knowledge-centered environment that is also learner-centered, new ideas would be introduced as they arise naturally in the course of teachers' activity – a "just-in-time" response to teachers' problems of practice. In sum, knowledge-centered professional learning environments:

- Draw on knowledge of accomplished teaching (*Design Conjecture 3*)
- Respond to issues that come up in teachers' ongoing instruction (*Design Conjecture 4*)

Centering Assessment and Feedback

Third, in learning environments that center assessment, learners get ample opportunities for feedback and revision as they work toward their goals. This is another place where even the best professional development activities often fall short. Even in workshops that provide teachers with opportunities to approximate authentic instruction (Grossman & McDonald, 2008) through planning or rehearsing instructional activities, teachers often do not get specific feedback on their work (Lynch et al., 2019). When they do, peer feedback does not always align with students' feedback: Even experienced colleagues, particularly if they teach in different school settings, cannot always predict the dynamics of particular classes (Kennedy, 2010). In the end, teachers are often left alone to interpret students' negative responses to novel teaching practices, leaving them with a challenging interpretive problem: Did the teacher do a poor job in implementing a new practice? Did students need more support in trying something new? Or does the practice not work? We note that well-developed approaches to professional learning, like video clubs (Sherin & Han, 2004) or lesson study (Lewis et al., 2006), include mechanisms for feedback, giving teachers opportunities to see and reflect on how instruction unfolds in real classrooms. In our vision of an assessment-centered teacher learning environment, assessments should reflect teachers' learning goals by revealing their current understandings of important ideas and provide information so they can adjust instruction accordingly to make progress toward their goals. Using language from Chapter 3, feedback is fodder for new interpretations that support revisions to teachers' pedagogical responsibilities and pedagogical actions, putting them in better alignment. Thus, assessment-centered professional learning environments:

- Provide adequate and timely feedback on teachers' attempts to improve their instructional practice to support their ongoing efforts (*Design Conjecture 5*)

Centering Teacher Community

Finally, learning environments that center community promote connections between learners' activities and broader communities. Sociological studies of schools have noted the correlation between higher-than-expected student outcomes and the presence of strong teacher communities (Lee & Smith, 1996; Ronfeldt et al., 2015). Strong teacher communities that are focused on student learning can support teacher learning and sustain improvement efforts. For this reason, a whole body of work seeks to identify the learning resources that exist within teacher communities (see Lefstein et al., 2020 for an overview). Ultimately, community-centered professional learning environments:

- Provide teachers with a community of like-minded colleagues to learn with and garner support from as they work through the challenges of ambitious instruction (*Design Conjecture 6*)

When these four lenses on learning environments are aligned through shared goals about what is taught, how it is taught and how it is assessed, learners are best positioned to develop new understandings.

Recontextualizing Instructional Practices

Bransford and colleagues' *How People Learn* framework is largely built upon the most extensively studied cases of learning: children in schools. We need to consider the nature of teacher knowledge – the subject-world relations summarized in Chapter 1. For instance, the Competing Visions of Quality and Interpretation Premises mean that instructional practices offered in professional development settings are likely to be transformed when brought into the classroom. Recall Kennedy's (2016) observations that prescriptive forms of professional development have little influence on instructional practice and student learning. Teacher learning is not a plug-and-play proposition, despite the pervasive language of implementation that surrounds instructional improvement efforts. As Fred Erickson (2014) reminds us, "a teacher does not teach children in general, but particular children in particular circumstances" (p. 3). To support meaningful learning, designs should account for those particulars. Thus, to better account for the active, interpretive work teachers do when they bring new practices in their classroom, we find *recontextualization* (Horn et al., 2013; van Oers, 1998) to be a more useful term.

The work of recontextualizing instructional practices involves its own set of practices – things teachers do to bridge the often-puzzling gaps between idealized instructional practices and their lived realities. In Samantha Marshall and Ilana Horn's (2021) analysis of teachers taking up ideas learned in professional development in their classrooms, we found that contextually situated goals shaped teachers' learning, and these goals did not always align with intended instructional practice. For instance, a teacher may learn an assessment practice designed to offer students feedback on their in-progress learning (Wiliam, 2011). The appeal of the practice to the teacher, however, comes from how it helps him meet another school-specific goal of *increasing passing rates*. While the two purposes are not inherently incompatible, one teacher's recontextualized version of the practice met the second goal more successfully than the first, as he offered students multiple revisions of tests without consistently supporting their meaningful learning. Recontextualization necessitates adaptation. However, some adaptations make the practice better fit the teacher's context while preserving its core purpose, and others do not.

Recontextualization challenges are inevitable as teachers incorporate new practices and modify their instruction. To imagine how these challenges might be addressed, we return to the conceptual resources for teachers' learning described in Chapter 2. In particular, we consider the role of representations of practice (Little, 2003), positing that *rich representations of practice that are close to teachers' own contexts minimize the translational and interpretive demands of recontextualization*. As teachers gather evidence to inform their pedagogical judgments, representational adequacy becomes consequential; when representations are not substantive or reliable, teachers lean on their trust in and admiration for the person offering feedback (as with Mr. Green in Chapter 3). Looking at video clubs and lesson study as examples, we note that these interventions include representations of practice that come from teachers' or their colleagues' classrooms, suggesting that this design feature might have added potency for conceptual change. This leads to our next design conjecture. To support teachers' conceptual change, professional learning activities should:

- Provide teachers with rich images of their own classroom teaching to minimize the burden of recontextualization (*Design Conjecture 7*)

Supporting Dialogic Learning Through Co-Inquiry into Practice

As often happens in design research, our pilot studies surfaced the need for an additional design conjecture that did not emerge from our gap analysis. As our team worked with teachers during our pilot year, we recognized the importance of developing a shared stance of supporting co-inquiry into practice. In many ways, this was an extension of the PDO leaders' commitments. At the same time, because of the novelty of our activity and its proximity to teachers' classrooms, it took a slightly different form in our work.

By supporting co-inquiry, we refer to being deliberately humble and uncertain in sharing of our interpretations of teachers' instruction and positioning our expertise as complementary to (not better than) the teachers'. Undoubtedly, our research team saw and noticed things we found interesting or important, but we also believed that the teachers knew their students, curriculum, and teaching contexts better than we ever could. Occasionally, some teachers did ask what we, as researchers, thought they should do in their classrooms – and, of course, we answered them, usually in the form of a brainstorm – but we did not presume to have all of the answers. This was not false humility on our part: Every member of our team had taught mathematics, and we understood that the teachers' interpretive resources differed from – but were not necessarily better or worse than – ours as researchers. Moreover, the co-inquiry stance was shared by each school-based team. The participating teachers often made suggestions to their colleagues, highlighting their different experiences and perspectives. In these ways, our design brought different kinds of expertise together as we made sense of the complex work of teaching (Horn, 2019; Jurow et al., 2019).

This set of commitments connects to theoretical notions of *hybridity* in learning (Gutiérrez et al., 1999; Gutiérrez, 2008), which have been described as ways learning environments can support new and expansive forms of knowledge by honoring the "onto-epistemic heterogeneity" (Warren et al., 2020, p. 278) within any group. Co-inquiry into practice is a hybrid activity – part research, part professional development, part collegial conversation – and, at their best, these conversations offered meaningful contexts for dialogue to derive new interpretations of and possibilities for instruction. The expansiveness inherent in this approach is crucial, as it contrasts with dominant onto-epistemics about teaching, particularly notions of "best practices." Warren and colleagues (2020) refer to such paradigms as "zero-point epistemologies," stances that privilege one set of understandings while "denying other perspectives, histories, and subjectivities" (p. 279). Since dialogue is at the heart of concept development (see Chapters 2 and 3), fostering authentic dialogue was an important goal for our work; our stance of co-inquiry and commitment to hybridity moved conversations away from the certainty of zero-point epistemology toward expansive forms of knowledge-making to support new understandings.

Thus, our final design conjecture was that, to support teachers' conceptual change, professional learning activities should:

- Respect teachers' autonomy, agency, and experiences as sensemakers by taking a stance of co-inquiry into instructional practice and foster interpretive dialogue (*Design Conjecture 8*).

From Conjectures to Professional Learning Design

Our conjecture mapping could inform any professional learning design, but we sought to support the PDO. To this end, we leveraged these conjectures about effective teacher learning environments to enhance the PDO's work supporting teachers' development of ambitious and equitable instruction (Table 4.1). Perhaps unsurprisingly, since the PDO's program reflected features of high-quality professional development, it also reflected many aspects

TABLE 4.1 A summary of design conjectures for teachers' professional learning environments.

Conjecture 1	Professional learning activities should address teachers' existing concepts about and practices for teaching.
Conjecture 2	Professional learning activities should align with teachers' personal goals for their learning.
Conjecture 3	Professional learning activities should draw on knowledge of accomplished teaching.
Conjecture 4	Professional learning activities should respond to issues that come up in teachers' ongoing instruction.
Conjecture 5	Professional learning activities should provide adequate and timely feedback on teachers' attempts to improve their instructional practice to support their ongoing efforts.
Conjecture 6	Professional learning activities should provide teachers with a community of like-minded colleagues to learn with and garner support from as they work through the challenges inevitable in transformative learning.
Conjecture 7	Professional learning activities should provide teachers with rich images of their own classroom teaching.
Conjecture 8	Professional learning activities should respect teachers' autonomy, agency, and experiences as sensemakers by taking a stance of co-inquiry into instructional practice and foster interpretive dialogue.

Note: Conjectures highlighted in white were addressed in the PDO's existing design. Conjectures highlighted in gray needed to be addressed through the partnership work.

of our design conjectures. For instance, when teachers identified a critical need in their school and proposed how they would work to improve it, this was consonant with Conjecture 2: *Professional learning goals need to align with teachers' personal learning goals.* In line with Conjecture 3, the PDO provided robust images of accomplished mathematics teaching during monthly workshops where ambitious instruction was modeled and discussed. Finally, fellows applied in school teams and learned alongside the PDO's network of ambitious mathematics educators. This aligned with Conjecture 6: *Teachers need a community of like-minded colleagues.* Nonetheless, our PDO partners believed that we could enhance their programs using the remaining design conjectures. For this reason, our co-design work centered on working on program improvement around the four remaining design conjectures, highlighted in gray in Table 4.1.

We noted that Conjecture 5 – about adequate and timely feedback – seemed to most closely echo the teachers' sentiments in the external evaluator's reports. Indeed, as we developed these conjectures, our PDO partners concurred that, at the start of our work together, teachers received inconsistent feedback about their teaching. They were ready and willing to work with us to develop formative feedback processes to support teachers' concept development more deliberately. For this reason, we focused our design on developing an activity that centers feedback, while addressing the other conjectures.

Because teachers entered the fellowship with their own learning goals, our design needed to be flexible to account for variation among learners, settings, and goals. Thus, we aimed to develop formative feedback processes that were close to the classroom (Conjecture 7) to enhance teachers' professional learning environments. By *formative feedback*, we refer to tools and processes that ascertain learners' current understanding and responsively adjust learning activities to better guide them toward their learning goals. These ideas are the heart of good formative feedback, and they align with the target design conjectures highlighted in gray in Table 4.1. While most descriptions of formative feedback center on students in classrooms, formative feedback can be adapted to teachers' professional learning, with adjustments for the sociocultural nature of professional knowledge in teaching.

To meet our various design goals, we developed *video-based formative feedback cycles* (VFFs, Figure 4.1) to support teachers' concept development around ambitious and equitable mathematics teaching. The VFF met our design conjectures by providing access to teachers' existing conceptions of their learning goals (Conjecture 1); responding to issues that arise in instruction (Conjecture 4); providing timely and adequate feedback to guide their ongoing efforts (Conjecture 5); providing teachers with rich images of their own classroom teaching (Conjecture 7); and supporting a co-inquiry stance to foster productive interpretive dialogue (Conjecture 8). Central to the VFF design is the notion that teachers focus on personally relevant learning goals through co-inquiry into their own practice.

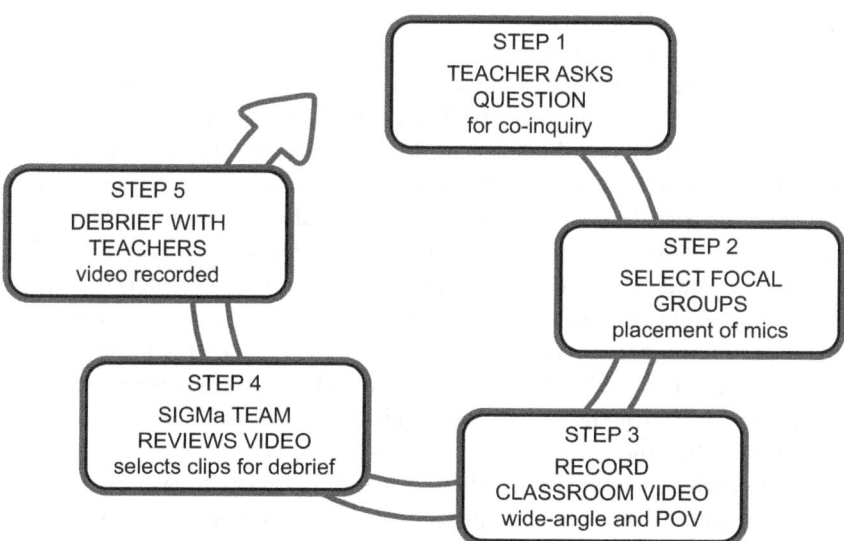

FIGURE 4.1 An overview of the video-formative feedback (VFF) process.

After iterative refinement during our pilot year, the VFF process ended up unfolding as follows: Before the lesson, we consulted with the focal teacher – often via email – to identify a co-inquiry question (Step 1 in Figure 4.1). Typically, we asked the teacher *what they were working on in their teaching*. Once a question was decided, we asked them to help us think about *how we can look for evidence* of how they were doing.

On the day of the lesson, we asked the teacher to select four focal student groups[2] to capture their dialogue during our recording (Step 2). Sometimes, teachers chose groups at random; other times, they chose students whose work could shed light onto their inquiry question. Then we used two cameras to document instruction (Step 3). One captured the teacher's point-of-view via a camera mounted on their shoulder. Another captured the whole-class activity through a camera mounted on a robot tripod (Swivl), which pivoted to track the teacher as they walked around the classroom. The Swivl recording also captured audio from microphones at the four focal groups' tables. We also documented lessons through fieldnotes and photographs of lesson artifacts (e.g., handouts or boardwork) to supplement the video recordings. At the end of the lesson, we checked in with the focal teacher to get their "gut check" on what happened during the lesson and note any concerns they had or moments they wanted us to look at closely.

Soon after the lesson, we reviewed the video, looking for interactions that supported the teacher's inquiry (Step 4). We also identified instructional moments that were tangential to the teacher's inquiry but that elucidated important aspects of instruction. Our video review strongly resembled the analytic work we would do if we were analyzing the lesson for a research paper, but we were guided by the teachers' curiosities. In this way, it was an authentic co-analysis with the teachers.

Typically a day or two after the lesson, we met with the focal teacher and their PDO partner teachers (Step 5). The debrief usually began with the focal teacher giving a recap of the lesson and their questions. Then we described the moments we selected for our shared review, sometimes suggesting we start with one in particular and sometimes asking for their input about where to begin. Since the lessons were recent, they usually had clear memories of the moments or students they were curious about. We usually introduced a clip – where it fell in the lesson sequence or why it was of interest, sometimes with an account of its typicality or uniqueness – viewed it together, and then asked the focal teacher how the clip helped them understand their question. The teachers reported that they particularly enjoyed seeing and hearing students' interactions when they, themselves, were not involved.

In the debriefs, we typically had moments of *high press*, where we engaged the focal teacher's conceptions of pedagogical issues and probed them. As we negotiated these conceptions with teachers and their colleagues, we analyzed them as *episodes of pedagogical reasoning (EPRs; Horn, 2005)* – moments where teachers' sensemaking is revealed, uncovered, and even contested. These EPRs illuminated teachers' interpretations of classroom moments and the concepts they invoked to make sense of these moments. Foreshadowing our findings, these EPRs became potent sites for concept development, offering rich representations of pedagogical action, inviting extended narratives of pedagogical reasoning that surfaced, engaged, and refined pedagogical responsibilities, informing new courses of pedagogical action. As teachers developed new interpretations of instructional moments, they imagined other ways of working in the future that better aligned with their commitments to ambitious and equitable mathematics teaching, reconceptualizing their work.

Studying Teachers' Concept Development

In Year 0 of Project SIGMa, we familiarized ourselves with the PDO as a learning environment, getting to know the teachers and their schools, shadowing them at the monthly workshops and professional conferences. In the latter half of the year, we also began piloting the VFF, refining our process through iterative design cycles (Cobb et al., 2003). At the end of Year 0, we recruited focal teams who committed to working with us for a two-year VFF intervention. Participants are summarized in Table 4.2. Participating teachers taught grades 6–12, had between 5 and 25 years of teaching experience (mean = 14.69 years), and participated in 3–11[3] VFFs throughout the project, with at least one as the focal teacher (mean = 6.08 VFFs). Given our team's visibility, we are deliberately vague about potentially identifiable details about our participants.

The intensive intervention and data collection for Project SIGMa took place in Years 1 and 2. Across the board, participating teachers were considered leaders in their schools, often serving as department chairs or on school-level leadership committees. Our participants came from two middle schools and three high schools.[4] Some of the schools' math departments, like Banneker's, were aligned with their vision of ambitious and equitable instruction; in other departments, this form of instruction was, at best, atypical and, at worst, contested. No matter the micropolitics of their departments, school administrators generally supported their efforts, since their PDO improvement projects signaled a commitment to the greater school community. They were thus recognized for their dedication to students and their investment in their own professional growth. Most schools primarily served students in their surrounding neighborhoods, although the district's open enrollment policy meant that students came from across the city. Their schools' sizes (rounded to the nearest multiple of 5 to prevent reverse lookups) ranged from 680 to 1850, with a mean of 1110 students. The schools' student demographic profiles ranged from 55% to 95% Latinx, 5% to 20% Asian, 15% to 20% Black, 5% to 10% White, 70% to 100% students qualifying for Free and Reduced–Price

TABLE 4.2 Participants in Project SIGMa.

School	Lead Researcher	Participant	Grades	Years Teaching[a]	VFFs Focal/Total	What They Learned
Falconer MS	Samantha (Sammie) Marshall	Lee Bellver	8th	15–20	5/11	Supporting students' conceptual understanding Supporting productive struggle Deciding what questions to ask
		Doha Arzoomanian	6th–8th	15–20	6/11	Supporting students' conceptual understanding Aligning instruction to the goal of understanding
Rees MS	Patricia (Patty) Buenrostro	Ezio Martín[b] (Y1)	8th	20–25	2/3	Scaffolding students' engagement in problem-solving tasks
		Veronica Kennedy (Y1)	8th	10–15	4/6[c]	Navigating social dynamics to foster productive student-to-student math talk
		Veronica Kennedy (Y2)	9th–10th			
Banneker HS	Grace Chen	Franck To	9th–12th	15–20	5/6[c]	Students' interaction with their peers Individual students' learning How to plan for peer interaction and individual learning
		Clark Zapatero	9th–12th	15–20	6/6[c]	Thinking about task design and student engagement with ideas "Leveling up" the quality of conversation about teaching

(*Continued*)

TABLE 4.2 (*Continued*)

School	Lead Researcher	Participant	Grades	Years Teaching[a]	VFFs Focal/Total	What They Learned
Noether HS	**Ilana (Lani) Horn & Nadav Ehrenfeld**	**Abigail Graham (Y2)**	9th–12th	10–15	4/6[c]	Strategically using groupwork structures to facilitate student discourse and conceptual understanding
		Abigail Graham (Y1)	9th–12th			Thinking more deeply about content
		Brad Miller	9th–12th	5–10	3/6	Setting up and supporting productive groupwork
		Greg Kahae	9th–12th	20–25	1/6	Setting up and supporting effective groupwork / Fostering student mathematical discourse / Considering social dynamics between students
Fermat HS	**Brette Garner**	*Marisa Dawson*[d]	*9th–12th*	*15–20*	*1/6*	
		Julie Woodman	9th–12th	5–10	3/6	Connecting instruction to students' understanding
		Lizette McLoughlin	9th–12th	5–10	3/6	Fostering productive groupwork to support students' conceptual understanding

[a] We report teachers' experience as a range to avoid sharing potentially identifiable details.

[b] Ezio chose not to continue participating in Year 2, when his colleague, Veronica, moved to Banneker.

[c] At Banneker, each teacher was observed and recorded by the research team, then participated in one team VFF debrief. For these debriefs, each teacher on the Banneker team was considered a focal teacher because we reviewed and incorporated data from each of their classrooms.

[d] Due to time constraints and other logistical factors, we were not able to collect sufficient data with Marisa to make strong claims about her learning.

Lunch, 10% to 35% native Spanish speakers, and 5% to 20% students with disabilities.

The teachers' standing as leaders in their schools greatly facilitated our project. Because of the intensive nature of our data collection, we needed both guardian consent and student assent for filming approximately 1200 students in focal teachers' classrooms. Additionally, our agreement with the district allowed us to exclude at most five students from being recorded, and then only if they received an "equivalent educational experience" like tutoring from a teaching assistant. Surprisingly, we managed to meet this high bar every time; we take this as indicative of students' and families' trust in their teachers.

In Years 1 and 2, we conducted regular VFF cycles with the focal teachers. Including the pilot VFFs in Year 0, we conducted a total of 35 VFFs. The VFF cycles alone offered rich, interconnected data for our analysis. We supplemented these with 75 interviews with focal teachers, observations of 115 sessions at 24 monthly PDO meetings, eight days of shadowing teachers at professional conferences, and various emails and informal conversations.

Our analysis focuses primarily on the VFF data, so we carefully documented these events. To that end, shortly after VFF debriefs, research team members also wrote structured memos, highlighting any notable aspects of the VFF, including technical difficulties (e.g., equipment malfunctions), interruptions (e.g., fire alarm), or unexpected insights. Additionally, each VFF debrief conversation was filmed and transcribed, giving us rich records of teachers' sensemaking around their classroom videos. As we describe in more detail in Chapter 5, we inductively coded debriefs for the *problems of practice* that emerged, with special attention to which speaker brought them up. This offered a window into teachers' engagement with different problems of practice over time.

After two years of our VFF intervention, we created learning profiles of each teacher, noting the problems of practice they frequently brought up. Connecting across the VFF debriefs they participated in – and supplementing with interviews and other data – we constructed *learning portraits* that captured their evolving understandings about teaching. At the end of Year 2, we conducted Exit Interviews with participants about their experiences with the VFF.

In a follow-up Member Check visit (Year 3), we developed brief summaries of each teacher's learning portrait to share with them. To look for evidence of sustained learning beyond the time of the intervention, we conducted two-person classroom observations, taking fieldnotes and paying particular attention to the themes they brought up in VFF debriefs. For example, our problems of practice inventory revealed that Veronica Kennedy frequently raised questions about *navigating social dynamics to foster productive student-to-student math talk*. When we observed her for our Member Check, we took careful note of her designs to foster student-to-student talk.

We then interviewed each teacher, sharing our written summary of their learning to see if they agreed with our assessment and using our recent classroom observations to illustrate or complicate our emerging findings. Although we explicitly invited them to, none of the teachers contradicted our narratives altogether. Seven out of 10 teachers agreed with our takes on their learning, while three modified our understanding, primarily by adding details we had not noted. For instance, we learned that Clark Zapatero's views on testing and assessment, which were not frequent topics of VFF debriefs (see Chapter 5), had been influenced by what he learned during his interactions with the research team within and beyond the VFFs:

> It didn't make it onto your guys' list. Maybe it was a me-and-Grace conversation. Grace said, "Why do you even give tests?" That one definitely stuck with me… That was definitely a good conversation I really remember from the process. There's another one that was about grading: "If you know where they're at and you've already assessed them, why are you giving them tests?" We talked about that for a while. And I remember that definitely sticking.
>
> *(Clark Zapatero, Exit Interview, May 2019)*

Clark told us that this conversation with Grace outside the VFFs, along with discussions of assessment with his colleagues during the VFFs, led him to introduce a test retake policy in his Algebra classes. We see this as an account of Clark's conceptual change about assessment: He revised pedagogical actions, reasoning, and responsibility, even though it was not richly captured in our data.

In the empirical chapters that follow, we illustrate how the VFF supported teachers' concept development toward ambitious and equitable mathematics instruction. In Chapter 5, we argue that the hybrid space of the VFF uniquely supported teachers' concept development about important aspects of this practice. We use an example from Abigail Graham's learning to illustrate how the VFF structure supported reconceptualizations of these problems in productive ways. Using the problems of practice coding of the VFF debriefs, we show the range of topics that teachers discussed. In Chapter 6, we show how teachers learned how to learn in VFFs, particularly as they renegotiated their subject-world relations and conceptions of good teaching. We draw on the learning portraits of Greg Kahae, Lizette McLoughlin, and Veronica Kennedy to show different contours of this renegotiation. In Chapter 7, we use the cases of Doha Arzoomanian, Ezio Martín, and Lee Bellver to show how teachers' learning from VFFs could sometimes come through brief, critical incidents that spurred them to refine their pedagogical judgment as they reconceptualized their practice. In contrast, in Chapter 8, we look at how teachers used the VFFs to learn over time, as was the case for Brad Miller and Julie Woodman.

We shared the penultimate drafts of these chapters with the teachers as a final member check, seeking corrections on factual errors and to identify possible interpretive conflicts between their experiences and our analysis.

Notes

1 Goals, of course, are not the property of individuals; teachers' goals for learning to teach, as we argued in Chapter 3, are shaped by their institutional and collegial contexts.
2 All recorded students were fully consented.
3 Samantha Marshall secured additional funding to experiment with other forms of VFF cycles with the Falconer team, resulting in their higher than typical number of VFFs.
4 Given the intensive nature of our video data collection, we could only film in classes with children aged 12 and up, limiting us to 8th grade and above.

References

Ball, D. L., & Cohen, D. (1999). Developing practice, developing practitioners: Toward a practice-based theory of professional education. In G. Sykes & L. Darling-Hammond (Eds.), *Teaching as a learning profession: Handbook of policy and practice* (pp. 3–32). San Francisco: Jossey-Bass.

Banilower, E. R., Smith, P. S., Malzahn, K. A., Plumley, C. L., Gordon, E. M., & Hayes, M. L. (2018). *Report of the 2018 NSSME+*. Chapel Hill, NC: Horizon Research, Inc.

Bean, R. M., Draper, J. A., Hall, V., Vandermolen, J., & Zigmond, N. (2010). Coaches and coaching in reading first schools: A reality check. *Elementary School Journal, 111*, 87–114.

Boaler, J., & Humphreys, C. (2005). *Connecting mathematical ideas: Middle school video cases to support teaching and learning*. Heinemann.

Bransford, J., Brown, A. L., & Cocking, R. R. (1999). *How people learn: Brain, mind, experience, and school*. National Academies Press.

Breault, R. (2009, October). The celluloid teacher. *The Educational Forum, 73*(4), 306–317.

Cobb, P., Confrey, J., diSessa, A., Lehrer, R., & Schauble, L. (2003). Design experiments in educational research. *Educational Researcher, 32*(1), 9–13.

Cohen, D. K. (1990). A revolution in one classroom: The case of Mrs. Oublier. *Educational Evaluation and Policy Analysis, 12*(3), 311–329.

Darling-Hammond, L., Hyler, M. E., & Gardner, M. (2017). Effective teacher professional development. Learning Policy Institute. https://learningpolicyinstitute.org/product/teacher-prof-dev

Desimone, L. M. (2009). Improving impact studies of teachers' professional development: Toward better conceptualizations and measures. *Educational researcher, 38*(3), 181–199.

Desimone, L. M., Porter, A. C., Garet, M. S., Yoon, K. S., & Birman, B. F. (2002). Effects of professional development on teachers' instruction: Results from a three-year longitudinal study. *Educational evaluation and policy analysis, 24*(2), 81–112.

Erickson, F. (2014). Scaling down: A modest proposal for practice-based policy research in teaching. *Education Policy Analysis Archives, 22*(0), 1–9. https://doi.org/10.14507/epaa.v22n9.2014.

Gallucci, C., Van Lare, M. D., Yoon, I. H., & Boatright, B. (2010). Instructional coaching building theory about the role and organizational support for professional learning. *American Educational Research Journal, 47*, 919–963.

Garet, M. S., Porter, A. C., Desimone, L., Birman, B. F., & Yoon, K. S. (2001). What makes professional development effective? Results from a national sample of teachers. *American Educational Research Journal, 38*(4), 915–945.

Goldsmith, L. T., Doerr, H. M., & Lewis, C. C. (2014). Mathematics teachers' learning: A conceptual framework and synthesis of research. *Journal of Mathematics Teacher Education, 17*(1), 5–36.

Grigg, J., Kelly, K. A., Gamoran, A., & Borman, G. D. (2013). Effects of two scientific inquiry professional development interventions on teaching practice. *Educational Evaluation and Policy Analysis, 35*(1), 38–56.

Grossman, P., & McDonald, M. (2008). Back to the future: Directions for research in teaching and teacher education. *American Educational Research Journal, 45*(1), 185–205.

Gutiérrez, K. (2008). Developing a Sociocritical Literacy in the Third Space. *Reading Research Quarterly, 43*(2), 148–164.

Hall, R., & Horn, I. S. (2012). Talk and conceptual change at work: Adequate representation and epistemic stance in a comparative analysis of statistical consulting and teacher workgroups. *Mind, Culture, and Activity, 19*(3), 240–258.

Henningsen, M., & Stein, M. K. (1997). Mathematical tasks and student cognition: Classroom-based factors that support and inhibit high-level mathematical thinking and reasoning. *Journal for Research in Mathematics Education, 28*(5), 524–549.

Horn, I. S. (2005). Learning on the job: A situated account of teacher learning in high school mathematics departments. *Cognition and Instruction, 23*(2), 207–236.

Horn, I. S. (2012). *Strength in numbers*. Reston, VA: National Council of Teachers.

Horn, I. S., Nolen, S. B. & Ward, C. J. (2013). *Recontextualizing practices: Situative methods for studying the development of motivation, identity and learning in and through multiple contexts over time.* In M. Vauras & S. Volet (Eds.), *Interpersonal regulation of learning and motivation: Methodological advances.* Routledge.

Horn, I. S. (2017). *Motivated: Designing math classrooms where students want to join in.* Heinemann.

Horn, I. S. (2019). Supporting the development of pedagogical judgment: Connecting instruction to contexts through classroom video with experienced mathematics teachers. In *International Handbook of Mathematics Teacher Education: Volume 3* (pp. 321–342). Brill Sense.

Horn, I., Garner, B., Chen, I. C., & Frank, K. A. (2020). Seeing colleagues as learning resources: The influence of mathematics teacher meetings on advice-seeking social networks. *AERA Open, 6*(2), 2332858420914898.

Horn, I. S., Garner, B., Kane, B. D., & Brasel, J. (2017). A taxonomy of instructional learning opportunities in teachers' workgroup conversations. *Journal of Teacher Education, 68*(1), 41–54.

Horn, I. S., & Kane, B. D. (2015). Opportunities for professional learning in mathematics teacher workgroup conversations: Relationships to instructional expertise. *Journal of the Learning Sciences, 24*(3), 373–418.

Horn, I. S., & Kane, B. D. (2019). What we mean when we talk about teaching: The limits of professional language and possibilities for professionalizing discourse in teachers' conversations. *Teachers College Record, 121*(4), n4.

Horn, I. S., & Little, J. W. (2010). Attending to problems of practice: Routines and resources for professional learning in teachers' workplace interactions. *American Educational Research Journal, 47*(1), 181–217.

Ingersoll, R. M., & May, H. (2012). The magnitude, destinations, and determinants of mathematics and science teacher turnover. *Educational Evaluation and Policy Analysis, 34*(4), 435–464.

Jackson, P. W. (1990). *Life in classrooms.* Teachers College Press.

Jackson, K., & Cobb, P. (2013). Coordinating professional development across contexts and role groups. In M. Evans (Ed.) *Teacher Education and Pedagogy: Theory, Policy and Practice,* pp. 80–99.

Jackson, K., Horn, I. S., & Cobb, P. (2018). Chapter four: Overview of the teacher learning subsystem. In P. Cobb, K. Jackson, E. Henrick, T. M. Smith, & MIST team (Eds.), *Systems for instructional improvement: Creating coherence from the classroom to the district office* (pp. 65–75). Harvard Education Press.

Jurow, S., Horn, I. S., & Philip, T. M. (2019). Re-mediating knowledge infrastructures: A site for innovation in teacher education. *Journal of Education for Teaching, 45*(1), 82–96.

Kane, B. D., & Rosenquist, B. (2019). Relationships between instructional coaches' time use and district-and school-level policies and expectations. *American Educational Research Journal, 56*(5), 1718–1768.

Kane, T. J., & Staiger, D. O. (2008). *Estimating teacher impacts on student achievement: An experimental evaluation (No. w14607).* National Bureau of Economic Research.

Kazemi, E., & Hubbard, A. (2008). New directions for the design and study of professional development: Attending to the coevolution of teachers' participation across contexts. *Journal of Teacher Education, 59*(5), 428–441.

Kennedy, M. M. (2010). Attribution error and the quest for teacher quality. *Educational Researcher, 39*(8), 591–598.

Kennedy, M. M. (2016). How does professional development improve teaching? *Review of Educational Research, 86*(4), 945–980.

Lampert, M., & Ball, D. L. (1998). *Teaching, multimedia, and mathematics: Investigations of real practice. The practitioner inquiry series.* Teachers College Press.

Lee, V. E., & Smith, J. (1996). Collective responsibility for learning and its effects on gains in achievement and engagement for early secondary school students. *American Journal of Education, 104*(2), 103–147.

Lefstein, A., Louie, N., Segal, A., & Becher, A. (2020). Taking stock of research on teacher collaborative discourse: Theory and method in a nascent field. *Teaching and Teacher Education, 88,* 102954.

Lewis, C., Perry, R., & Murata, A. (2006). How should research contribute to instructional improvement? The case of lesson study. *Educational Researcher, 35*(3), 3–14.

Little, J. W. (2003). Inside teacher community: Representations of classroom practice. *Teachers College Record, 105*(6), 913–945.

Lortie, D. C. (1976/2020). *Schoolteacher: A Sociological Study.* University of Chicago Press.

Lynch, K., Hill, H. C., Gonzalez, K. E., & Pollard, C. (2019). Strengthening the research base that informs STEM instructional improvement efforts: A meta-analysis. *Educational Evaluation and Policy Analysis, 41*(3), 260–293.

Muhlenbruck, L., Cooper, H., Nye, B. & Lindsay, J. (2000). Homework and achievement: Explaining different strengths of relation at the elementary and secondary levels. *Social Psychology of Education, 3,* 295–317.

Munson, J. (2018). *In the moment: Conferring in the elementary math classroom.* Heinemann.

Poglinco, S. M., Bach, A., Hovde, K., Rosenblum, S., Saunders, M., & Supovitz, J. A. (2003). *The heart of the matter: The coaching model in America's choice schools.* CPRE Research Reports. Retrieved from https://repository.upenn.edu/cpre_researchreports/35

Porter, A. C., Garet, M. S., Desimone, L. M., & Birman, B. F. (2003). Providing effective professional development: Lessons from the Eisenhower program. *Science Educator, 12*(1), 23.

Ronfeldt, M., Farmer, S. O., McQueen, K., & Grissom, J. A. (2015). Teacher collaboration in instructional teams and student achievement. *American Educational Research Journal, 52*(3), 475–514.

Sackstein, S. (2015, October). Differentiate PD for Optimal Teacher Engagement. *Education Week.* https://www.edweek.org/leadership/opinion-differentiate-pd-for-optimal-teacher-engagement/2015/10.

Sandoval, W. (2014). Conjecture mapping: An approach to systematic educational design research. *Journal of the Learning Sciences, 23*(1), 18–36.

Sandoval, W. A., & Bell, P. (2004). Design-based research methods for studying learning in context: Introduction. *Educational Psychologist, 39*(4), 199–201.

Schneeberger McGugan, K., Horn, I. S., Garner, B., & Marshall, S. (2021). 'Even when it was hard, you pushed us to improve': Negative emotions and teacher learning in coaching conversations [Manuscript submitted for publication]. Department of Teaching, Learning, and Diversity, Vanderbilt University.

Schneider, J. (2014). *From the Ivory Tower to the schoolhouse: How scholarship becomes common knowledge in education.* Harvard Education Press

Schoenfeld, A. H. (1988). When good teaching leads to bad results: The disasters of 'well-taught' mathematics courses. *Educational Psychologist, 23*(2), 145–166.

Sherin, M. G., & Han, S. Y. (2004). Teacher learning in the context of a video club. *Teaching and Teacher Education, 20*(2), 163–183.

Smith, M. S., & Stein, M. K. (2011). *5 practices for orchestrating productive mathematics discussions.* Reston, VA: National Council of Teachers of Mathematics.

Van Oers, B. (1998). From context to contextualizing. *Learning and Instruction, 8*(6), 473–488.

Warren, B., Vossoughi, S., & Rosebery, A. S. (2020). Multiple ways of knowing★: Re-imagining disciplinary learning. In N. S. Nasir, C. D. Lee, R. Pea, & M. M. de Royston (Eds.), *Handbook of the cultural foundations of learning* (1st ed., pp. 277–294). Routledge.

Wei, R. C., Darling-Hammond, L., & Adamson, F. (2010). Professional development in the United States: Trends and challenges. Dallas, TX. National Staff Development Council.

Wiliam, D. (2011). *Embedded formative assessment.* Solution Tree Press.

Wilson, S. M., & Berne, J. (1999). Chapter 6: Teacher learning and the acquisition of professional knowledge: An examination of research on contemporary professional development. *Review of Research in Education, 24*(1), 173–209.

Zembylas, M. (2003). Caring for teacher emotion: Reflections on teacher self-development. *Studies in Philosophy and Education, 22*(2), 103–125.

5
PUTTING FORMATIVE FEEDBACK INTO PRACTICE

With Patricia Buenrostro and Samantha Marshall

In this chapter, we give an overview of what Project SIGMa teachers encountered when they participated in video-formative feedback (VFF) cycles. It is one thing to design an intervention; it is another thing entirely to put it into practice. We describe the double edge of the rich representations of instruction that grounded our dialogue, with the increased visibility of the details of practice also bringing increased teacher vulnerability. We also articulate how we, as researchers, coordinated our facilitation practices, drawing on prior research on teacher learning and our pilot work. Then we illustrate the dialogism of our VFF facilitation through a close look at one instance of a teacher revising her inquiry questions, which we see as microgenetic conceptual change. Finally, we summarize the problems of practice that VFFs elicited, both from participating teachers and members of the research team, with an eye on what this suggests about learning about ambitious and equitable mathematics teaching.

The Vulnerability of Visibility: The Double Edge of Rich Representations of Practice

As we described in the Asynchronicity of Reflection Premise (Chapter 1), an endemic challenge to supporting teacher learning is that conversations about teaching typically happen apart from instruction. That is, opportunities to talk about practice – professional development workshops, collaborative planning sessions, and so on – are almost always separate from actual work with students.[1] As scholars have noted, this gap between action and sensemaking makes it difficult for professional development and teacher education courses to directly impact instruction (Horn & Campbell, 2015; Zeichner, 2010).

DOI: 10.4324/9781003182214-7

Especially when we take a perspective that insists on the irreducibly situative nature of teaching, the multiple, simultaneous details of classroom life shape discussions of what practices make sense and why they make sense.

Without synchronous examinations of teaching, conversations about teaching rely on representations of teaching practice (Little, 2003) – including lesson plans, student work, verbal replays of classroom interactions, or rehearsals for future instruction – that necessarily reduce the complex contexts they come from. As we have laid out in previous chapters, when sharing teaching problems with colleagues, it is hard to know which details are most relevant. Did the lesson fall flat because the teacher introduced a new collaboration format? Or was it because the class was before lunch, so students were hungry and tired? Comparing collaborative sensemaking across professionals in different disciplines, Hall and Horn (2012) referred to this question about which details matter for grounding interpretations as an issue of *representational adequacy*. For teachers, this includes identifying the right features of a problem to communicate when consulting with colleagues, since these help interlocutors identify, elaborate, and stabilize their accounts of what happened and troubleshoot how it might be improved. Often, issues of representational adequacy limit the depth and specificity of conversations about teaching. "The lesson bombed," a teacher might tell colleagues. "Oh, that's too bad. My students loved it," the reply might come. Aside from its lack of empathy, there is no representational traction in this exchange for making sense of these different outcomes. The lesson did not work for one teacher and worked for another, but without a fuller picture of what transpired, neither can make sense of their different experiences.

As we described in Chapter 4, the Project SIGMa team developed VFFs with issues of representational adequacy in mind. To create rich representations of instruction, we used multiple cameras (one whole-class view, another teacher point-of-view) and audio recordings of students' talk to ground debrief conversations (Conjecture 7). By selecting and sharing clips relevant to teachers' questions, we anchored our discussions in vivid details of practice, with the added ability to rewind, transcribe, explore, and compare instructional moments from different perspectives. Furthermore, watching video clips *with* teachers – rather than recounting what we saw (filtered through our lenses and biases) – allowed them to make their own sense of interactions. Indeed, in interviews, teachers told us rather frankly that the video was inherently more trustworthy than we would have been as observers. As Julie Woodman described:

> Seeing it come out of the students' mouths […] I just think it's more solid evidence of [an issue] and you don't actually need to say, "This is happening." […] Taking notes would be fine, too, but I just think the video adds. When you see your students – and I love my students so much – so

you feel when you see a blind spot, you're like, "Oh, crap. That's something I want to work on." [...] We can't be omniscient.

(Exit Interview, May 2019)

Other participants concurred, adding that the richness of videos often surfaced issues that they were not previously attuned to. "You see a lot of other things that you're not looking for when you watch those videos," Lee Bellver explained *(Exit Interview, June 2019)*.

Sometimes, teachers found that the videos affirmed their classroom designs, as Franck To recounted:

I'm very thoughtful about the smoothness of how my day goes. So, when do I pass out the next task? Little things like that. But I see that on the video, and I was like, "Okay, that didn't seem clunky!" [...] I was like, "Cool! The time I had planned for thinking [about] things, it's showing."

(Exit Interview, May 2019)

The richness of the video and audio representations, which we allowed teachers to review independently outside of debriefs, offered abundant resources for teachers' sensemaking about instruction.

In addition to developing rich representations of instruction, we designed VFFs to be hybrid learning spaces that elicit teachers' sensemaking, valuing their experiences and expertise alongside our own (Conjecture 8). We gave teachers feedback on their instruction with unusual specificity and depth, uncovering issues that we – or they – might not have otherwise considered, while addressing the issues they found most pressing (Conjectures 4 and 5). However, this increased visibility was accompanied by increased vulnerability. This additional risk was the double edge of the rich representations of practice.

Managing the Double Edge of Rich Representations: The Centrality of Trust

In line with Conjecture 8, the VFF design aimed to bring together both researchers' and teachers' expertise in non-hierarchical ways to support mutual learning (see also Jurow et al., 2019). Members of the research team all had experience in mathematics teaching and teacher education, but we also recognized that our teachers were accomplished and experienced in their own right. They were dedicated professionals with relatively long careers in the classroom. Beyond their professional knowledge, they were also experts in their own teaching contexts, knowing important details about their schools, communities, students, and state and district policies.

The co-inquiry aspect of our collaboration was sincere: The rich rep-resentations of practice we produced often surfaced instructional issues that were new to us in some important way, if only in their particularity. In taking this stance, we sought to eschew outside-in models of professional devel-opment, instead moving toward authentic partnerships with teachers. This paradigm shift in professional development offers a more democratic and inclusive way of working with teachers than is typical. For this reason, partic-ipants repeatedly pointed to the importance of trust in our work together. As Abigail Graham told us:

> One thing that makes this process so special is that I really trust all of you completely to pick out things that will be interesting and to do it in a way that will be helpful. I think in the future, I think if they're[2] looking to continue this process, they're going to have to be really careful what the goals are and how to achieve those goals and how to establish that trust.
>
> *(Exit Interview, May 2019)*

Affirmations like Abigail's suggested that we met the goal of establishing trust with the teachers, something crucial to our work. As we described in Chapter 4, we designed our intervention to foster trust by assigning each school team a primary research contact who facilitated their VFFs. By offering teachers a continuous relationship, we hoped to earn their trust and support their vulnerability in sharing their practice.

Of course, the need for trust went beyond individual researcher-teacher relationships. Engaging in VFFs required trust within the school teams them-selves. Veronica Kennedy described it aptly:

> It just is a vulnerable thing. Especially to show [the video] in front of your peers. I think you guys [the researchers] came in and you did the video, you watch the video, you analyze the video, you come back, you watch the video with me, that's one layer less risky than watching the video with everybody. That's part of it too. Your team has to be really strong. I have to be willing to be like, "Oh, I'm going to need your support, this is hard."
>
> *(Exit Interview, May 2019)*

Veronica's account highlights the double edge of the rich classroom records at the heart of the VFFs. On one hand, they afforded joint sensemaking about important instructional issues. On the other hand, that richness – what Lee called the "things you're not looking for" – removes the possibility of deflect-ing uncomfortable feedback by suggesting someone's critical interpretation comes from a difference in memory or perspective. Teachers never knew how lessons would go before filming, increasing their risk. Adding even more uncertainty to the mix, our team documented lesson details outside the teachers' direct observation. Lizette McLoughlin narrated her internal monologue when being filmed as, "There's the [feeling of], 'Oh my God, what did they just see? What could've possibly happened on those [student] microphones that I won't know about?'" (*Exit Interview, June 2019*). The possibility of the unexpected, the decreased room for plausible deniability in interpreting uncomfortable moments, and the exposure of backstage details made the VFFs quite risky for the teachers, requiring trust all around.

For these reasons, we would be naïve to overlook where the risk ultimately lay in our partnership during VFF encounters or to imagine that we could truly level the vulnerability we each brought to our co-inquiry. As researchers, we analyzed our own practices during VFF debriefs, which invited a layer of critique: *Why didn't I follow up on that comment? Did I press too hard on this issue? Was that connection clear?* In some ways, this paralleled the vulnerability that teachers felt – exposing both our facilitation and teachers' instruction to critique. However, these conversations took place primarily within our research team; of course, when we were concerned that we pushed too far or missed in some way, we would check in with the teachers for feedback. Nonetheless, despite our democratic intentions, we could not equalize the inherently uneven power dynamics that were a part of the VFFs. Ultimately, we produced durable records of teaching and conversations about teaching; while we shared them with our participating teachers, the records remained in our control. Clark Zapatero described the imbalance in social risk coming from this uneven power dynamic:

> I mean, you have videotape, you know, and audio tape. I think there was one thing I wanted scratched from the record. [*Brette: What do you mean?*] Just, I mean, that's intimidating. I mean, you have *all* that of me, and you could take it back – I don't think you are – but you could take that back and be evaluative behind closed doors if you wanted to. I don't think you are, but we have to trust that.
>
> (*Exit Interview, May 2019*)

Undoubtedly, the teachers whose instruction was being examined were inherently vulnerable in the VFF process, an issue we tried to attend to as a research team. (See Chapter 6 for more on this.)

Developing a Shared Stance on Facilitation: Fostering Teacher Learning Opportunities

To support a productive comparative analysis of participants' learning, we needed to lead VFFs in similar ways. This created new tensions in our research. On one hand, we needed to develop trusting relationships with teachers, which we supported by having a stable contact person with each team. On the other hand, we needed to develop shared approaches to facilitation, despite the differences in our team members' expertise, interactional styles, and personal and social histories. To manage this tension, we developed common approaches to facilitation in two ways. First, we reviewed our facilitation work in weekly team meetings, especially for the pilot VFFs (Year 0). Like the teachers discussing their practice, these reviews sometimes involved sharing stories, but we also watched and discussed video of our facilitation to learn from each other and coordinate our practices. Second, we grounded our work in prior research on supporting teacher learning through concept development in collegial conversations (see Chapter 3). We discussed these ideas informally throughout the project, and we distill them here as guiding principles for facilitation.

Guiding Principle 1: Identify and Build on Teachers' Strengths

All our participants brought significant expertise to the project. Examples of strengths included strong curricular knowledge, good relationships with students, well-organized lessons, and clear classroom routines. In practice, while we collected interesting classroom moments, we also worked to identify teachers' strengths – an asset orientation to teacher development. In short, an asset orientation assumes that people grow and learn best when they build on existing strengths. Motivationally, an asset orientation helps mitigate teachers' vulnerability by acknowledging what is working in their classrooms; we often started VFF debriefs with positive examples. Educationally, this ensured that we collectively acknowledged what went well in teachers' lessons. Conceptually, it helped anchor the brainstorms we often had in VFF debriefs. Because different aspects of teaching are interrelated, knowing teachers' strengths ensured that we did not propose something that would undo existing successes.

Guiding Principle 2: Support Teachers' Learning Opportunities Through Collective Interpretation

To support teachers' learning opportunities, we drew heavily on an earlier study where we analyzed over 100 teacher workgroups meetings to identify learning opportunities – essentially, to distinguish conversations that support teacher learning from those that do not (Horn et al., 2017). In that analysis, we operationalized learning opportunities by looking at teachers' interactions along two dimensions: (1) how they mobilized participants for future work (e.g., planning instruction) and (2) how they developed pedagogical concepts in conversation (e.g., explaining why an activity is productive). The richest learning opportunities occurred when groups developed concepts in conversation – discussing the *whys* of instructional strategies, not just *what* or *how*. We referred to these as *collective interpretation* meetings. Yet we found that, even in our purposive sample of "high-functioning" workgroups, collective interpretation was rare; teachers were more likely to decide *what* to do in upcoming lessons than to think about *why* they should (or should not) do those things. In VFF debriefs, we sought to facilitate collective interpretation conversations. As we discussed teachers' problems of practice (more on that below), we pressed them to consider the *whys*, in addition to the *whats* and the *hows*. Essentially, we talked about how to improve teaching, drawing on and building conceptual resources in the process.

Guiding Principle 3: Help Teachers Refine Their Pedagogical Judgment

As facilitators, we aimed to cultivate teachers' learning opportunities. Given our theoretical perspective on how that learning works – pedagogical action held up against pedagogical responsibility through pedagogical reasoning to sharpen pedagogical judgment (Chapter 3) – we strived to build connections across these. By eliciting teachers' pedagogical reasoning during VFFs, we could actively listen for their pedagogical responsibilities and hold them up against the pedagogical actions captured on video. This inevitably led to discussions of purpose – the *whys* of teaching – along with their understandings of how things unfold within any given lesson, the narrativized actions that constitute concepts.

Specifically, we used rich representations of practice to help teachers articulate, interrogate, and revise their choices. VFF debriefs did not always involve critique but aimed toward conceptual clarification by eliciting and discussing pedagogical reasoning. Sometimes, discussions involved unpacking why an interaction seemed to meet a teacher's goal in-the-moment (their pedagogical responsibility). At other times, we brainstormed alternative ways to respond – different pedagogical actions – and considered the potential tradeoffs in those choices.

Guiding Principle 4: Focus on Problem Framing and Concept Integration

Finally, we drew on Nicole Bannister's (2015) description of teachers' *problem framing* – how they diagnose and address problems of practice – as indicative of teachers' concepts about instructional issues. Importantly, Bannister found that problem frames shift as teachers develop greater facility with ambitious and equitable mathematics instruction, a finding supported by Jackson and colleagues' (2017) research on a larger sample of teachers. In Bannister's terms, as teachers became more proficient in ambitious instruction, their problem framings shifted from inactionable to actionable diagnoses of problems of practice. For example, instead of seeing a disengaged student as "flaky" (an inactionable framing), a teacher might recognize that the student is struggling to figure out classroom routines. This new diagnosis activates teacher agency – the possibility of addressing issues that arise, such as supporting the student in learning classroom routines – which is important for developing responsive teaching practices.

Drawing on these insights, we listened for teachers' problem framings and held them up against the classroom records to support their sensemaking. For instance, as we describe more fully in Chapter 8, Julie Woodman initially diagnosed a less successful lesson launch as stemming from students' difficulties with calculating the probability of independent events. After reviewing several video clips and discussing with the debrief team, she came to a new diagnosis: Details of pacing and activity design made it difficult for students to find probabilities quickly and apply them to the problem. This reconceptualization of students' difficulty is an example of a shift in diagnostic problem framing – one that offers a meaningful course of action.

We also pressed to integrate concepts to frame problems of practice since integrated concepts are a hallmark of accomplished, responsive teaching (Chapter 2). When Horn & Kane (2015) compared discussions in teacher workgroups with different levels of instructional accomplishment, we found that, among teachers whose practice most reflected ambitious and equitable instruction, teaching problems were frequently framed around the interrelationships between students' learning or experiences, instructional design, and mathematical content, which signaled integrated concepts. With this in mind, we often invited teachers to consider these connections as they made sense of video clips. In other words, if their initial framing emphasized a student learning issue (e.g., Julie's diagnosis that probability rules are hard to remember), we would encourage them to consider other underlying issues as well (e.g., Julie's revised diagnosis that included instructional design and pacing).

Ultimately, these guiding principles for facilitation had many overlaps. Facilitating for collective interpretation involves eliciting teachers' pedagogical judgment, which uncovers (and perhaps supports the revision of) problem framings. Sharing these ideas as a team helped us make in-the-moment decisions and coordinate our work across VFFs in different school contexts. Although important differences remained in our individual approaches to facilitation and our relationships with participants, enough commonality existed across VFFs for them to constitute a particular type of teacher-researcher encounter. Through our facilitation, we sought to invite teachers' thinking and put it in dialogue with both the video and the rest of the debrief team. As we illustrate next, this dialogism grounded in a shared vision of practice – critical in any hybrid learning environment (see Chapter 4) – was key to supporting teachers' reconceptualizations of problems of practice.

Dialogism in Action: Abigail Graham's Shifting Inquiry

To illustrate how teachers reconceptualized their practices in VFFs, we share how Abigail Graham's co-inquiry question evolved during one VFF debrief. We identify the questions that arose through the co-inquiry process and show how the emergent questions signal new understandings – a microgenetic, collective reconceptualization of Abigail's problem of practice. The evolving inquiry shows that, as the co-inquiry became more focused throughout the conversation, new questions introduced new problem frames. These frames offered more actionable concepts and activated Abigail's agency as a teacher.

Background and Context of Noether VFF 2

Abigail Graham taught at Noether High School during Year 1 of our project and then moved to Banneker during Years 2 and 3. She taught Algebra 1, Algebra 2, and Computer Science. From the beginning of our partnership, we noticed that Abigail was a highly reflective teacher. She often took notes during VFF debriefs, her brow furrowed in concentration. In an interview, she described herself as contemplative:

> It takes me a while to process things and wrap my head around things, and I would like the opportunity to talk about it and then have some initial questions get brought up, and then think about it more and come back to it.
> *(Member Check Interview, March 2020)*

She was comfortable with critique, having a background in a field that involved substantial public feedback, so she did not hesitate with that aspect of VFFs.

Abigail participated in six VFFs, including four as focal teacher. During VFF debriefs, many of the problems of practice that she raised fell into two themes: First, she posed questions about students' small-group discourse. Second, she raised issues about students' understanding of mathematics.

Focal Lesson: Synthetic Division

For the Noether VFF 2 debrief, Abigail taught an Algebra 2 lesson about synthetic division. She and a colleague, Greg Kahae, had co-planned the lesson, which involved a task called "Finding All the Roots." Abigail and Greg wanted to develop an inquiry activity around a boring topic: factoring quartic equations using synthetic division. Abigail began the class with the following warm-up:

For each function below:

a. Find ALL the zeroes.
b. Describe the end behavior.

c.   Sketch a graph of the function

1. $f(x) = -x^2 + 2x + 2$
2. $g(x) = 2x^3 - 8x^2 + 15x - 27$
3. $h(x) = x^4 - x^3 - 2x^2 + 6x - 4$

After students worked independently for 10 minutes, Abigail directed them to discuss the problems in pairs. After about five minutes, Abigail brought the class together and asked what made problem three "tricky." Students responded that they had not learned how to factor quartics. Abigail asked them to brainstorm strategies to solve fourth-degree polynomials. One student observed that $(x + 2)$ is a factor of the third function, explaining he had used synthetic division to identify that zero. Abigail asked students to discuss this idea; they worked through the reduced quartic together, using their knowledge of cubic equations to find the other factors. Then students, working in pairs, solved other quartics using this strategy.

Debrief Question 1: What are Students Talking About?

Before the lesson, Abigail shared her co-inquiry question with us. She asked us to pay attention to "student-student talk." She explained that she had incorporated some routines in her classroom to help facilitate students' mathematical conversations, and she wanted to know how effective they were. Specifically, she asked:

- What are students talking about during partner time?
- How can I better facilitate student-student talk so that they have a chance to explore ideas verbally?
- How can I better facilitate how students explain procedures?

(Email, January 2018)

As we planned the VFF debrief, we noted that most quartic equations could not be solved using this method, so we wondered what general strategies students might take away from this lesson. In our facilitation, we wanted to balance our question with Abigail's co-inquiry questions.

We began the debrief by orienting Abigail's colleagues – Brad Miller, Greg Kahae, and Marisa Dawson – to Abigail's lesson. Lani,[3] as lead facilitator, summarized Abigail's co-inquiry question and the flow of the lesson. She asked Abigail to describe her instructional goals, a prompt for her to articulate the pedagogical responsibility relevant to this lesson. Abigail and Greg described the lesson as supporting students' inquiry into solving higher-order polynomials – developing a procedure and understanding why it works.

Emergent Debrief Question 2: Would Students Benefit from More Time?

Throughout the VFF, we saw the questions Abigail raised as indicative of her own conceptual change about the issues. About 22 minutes into the debrief, we watched two video clips of student pairs (Corvin and Diego, then Jacinta and Kaleel[4]) working on the task. The student-to-student talk remained fairly procedural, despite Abigail's prompts to discuss strategy. After reviewing and discussing the video, Abigail said:

> Can I ask a question, to interject? I also am wondering about the timing of this. What I try to do is kind of feel out – like I have a little stopwatch – and I try to feel out how long I've been going. And I look at certain groups to see if they're finished, if they're not doing anything. But I'm also wondering, for that particular question, it seems like I didn't give them as much time as I had thought I had. *Would they have benefited from more time?* 'Cause I'm cutting them off right as Kaleel's explaining something to Jacinta that she might have questions about. But then it's such a tricky thing – timing – because it's hard to get them to use time to explore when they think they've already figured it out.

In this query, Abigail replayed her strategies for pacing groupwork and whole-class discussions: She tracked the time with a stopwatch, and she looked to see when certain students were finished. These additional details about how she shaped her pedagogical judgments gave us more sensemaking resources for our collective interpretation of the video. Abigail shared these strategies because viewing the video made her question how well they supported her goals ("it seems like I didn't give them as much time as I had thought I had"); she was comparing her pedagogical action to her pedagogical responsibility.

Abigail then wondered if additional time would have supported the kind of talk she hoped to see, offering a new problem framing. She said:

> Maybe what I need to do is be more specific about directing them to explain why, once they figure out what strategy they're gonna use. Because I think for both Jacinta and Kaleel, what probably would have happened if I had given them more time is they would've been like, "Oh, okay." And then same thing for Corvin and Diego.

In this statement, Abigail suggested that additional time would not necessarily have directed students toward more conceptual talk; she considered whether a specific directive to explain why their approach worked would better meet her goal. Lani agreed that the video showed that student talk stayed mostly procedural, supporting Abigail's diagnosis of the relationship between her pedagogical actions and responsibilities. Abigail expanded on the dilemma:

It's really hard to know, if you stop talking as a teacher, your hope is that somebody will keep thinking. And even if they're not talking at the moment, pretty soon they'll be like, "Hey what if..." you know, and keep going with it. But the times that it doesn't happen, then it's like those are wasted minutes that we could have done something with, you know? [*Lani: I know. Especially with yours*[5]] And so how do you find the balance? Because you don't know if they're gonna keep thinking about it or if they're gonna say, "So anyway, did you see that thing at lunch?"

Abigail's elaboration highlighted a keenly felt dilemma: whether giving students more time invites mathematical exploration or socializing. We confirmed that her concern was warranted, explaining that one focal group "finished" the task and sat around waiting for the rest of the lesson.

Emergent Debrief Question 3: What do Students Understand About What a Zero Is?

To push Abigail's sensemaking about her dilemma around fostering students' conceptual talk, Lani made a strong facilitation move, reframing the problem once again. She asked, "Can we get a little into the math, then? Like about what sense the kids are making?" Thus far, the group's collective interpretation emphasized the quality of students' mathematical talk – was it procedural or conceptual? – but had not specified how that might be discerned in this particular task. Abigail and the others agreed that would be worth discussing, so we shared clips that showed the details of students' sensemaking.

The first clip showed students factoring one of the quartic equations. In their factorization, the students found an imaginary zero.[6] One student said, "See, you need an *i*, right? It's not a zero, so no zero, *i* is not a zero. Which means there are no zeroes there." After viewing this clip, Brad jumped on the issue of imaginary zeroes: "Technically an *i*, you know, is a zero – but it's not a zero on the graph. So that's, like, a good, rich discussion." Lani pointed out that the second student responded to the first by saying, "Okay," signaling that the rich discussion did not happen.

Lani continued to press on the connection between conceptual and procedural talk by inviting the teachers to consider the students' understanding with the question: "What are they understanding about what a zero is? 'Cause there's the algebra and then there's the graph. And what connections are they making between the two?" By asking them to reflect on Abigail's teaching through the specific details surfaced in the video, this facilitation move pressed teachers toward both collective interpretation and an integration of concepts, linking student talk to Abigail's interest in their understanding of polynomials.

Because one conversation in one group does not represent an entire classroom's understanding, we played another clip of different students working on the same problem. One student named all four zeroes – real and imaginary – and said, "They're all zeroes but two of them you can't draw." Although the student did not get into the "good, rich discussion" that Brad had hoped for, the clip offered evidence that at least one student had sorted out that an imaginary root can exist, even if it does not show up on the graph.

Emergent Debrief Question 4: What is the Benefit of Discussing Imaginary Zeroes?

This discussion prompted another refinement in Abigail's question. She asked, "What benefit could there be from even discussing the imaginary zeroes – what richer conversations could we get by talking about the complex, the imaginary zeroes?" This launched a brainstorm about foregrounding a graphing strategy to build cross-representational connections, the limits of choosing the small subset of quartic equations that can be solved by factoring, and the dangers of glossing over the multiplicity of roots.

As the brainstorm progressed, Abigail introduced another question:

ABIGAIL: I guess, here's my question: So let's say they did have the graphs in front of them. Yeah? And then they're using that – like Brad, your idea was to say, "Here's the graph. If there are integer zeroes" – which I'm assuming there would be, like, one integer –

BRAD: At least one, yeah. I don't think I ever did a problem with none.

ABIGAIL: Right.

LANI: 'Cause there'd be no place to start.

BRAD: Well then you'd just say, you know, you can't use any of these techniques, like factoring or –

ABIGAIL: Right. "Prove to me that it is, in fact, a zero." So when they're dealing with the imaginary solutions from that, what would be the next part of that conversation? What would be the question to ask them? Or what would I want them to understand about the imaginary zeroes if they can't see them on the graph?

In brainstorming possible alternatives, Abigail searched for other ways of working in the future, including asking students to evaluate if certain numbers (real or imaginary) counted as zeroes, signaling that this was a learning opportunity for her. After a short discussion, Abigail continued:

So then, my next question is, like, what info … I don't know. And I think for myself, I don't know the answer. So, I'm also asking you guys. What would be interesting to them? What could they get out of these algebraic and graphical connections in terms of the imaginary zeros?

This led to additional questions about why to teach this topic at all, and whether it was included in the state standards or just in the textbook. Since we had anticipated this issue, we investigated these questions before the debrief. We shared that it was not listed in state standards as something students needed to know. In our theoretical terms, the teachers had no pedagogical responsibility to a policy or curricular document that should shape their instructional decisions on this; in fact, they may have more latitude than they had imagined regarding teaching students to factor quartics.

Emergent Debrief Question 5: What Should Students Understand and Say About Zeroes?

Lani then raised a broader question related to Abigail's goal of teaching mathematics for understanding:

> What would you say if you could name and articulate what kind of understanding you want the students to have at the end of Algebra 2, or at the end of this unit maybe, about what the zeroes of polynomials are? What do you want them to be able to say?

With this question, Lani sought to elicit an image of student learning and to refine Abigail's pedagogical responsibility in teaching this topic. If Abigail was not obligated to teach imaginary zeroes, was there a greater mathematical idea she wanted to help students grasp? After some deliberation, Abigail responded:

ABIGAIL: So, I want them to … and this is what we're testing them on, basically. It's like, given a polynomial function, can you find all of the real zeroes? That's what I really want them to know. And then given those real zeroes, can you make a rough approximation of what the graph is gonna look like?

LANI: Pretty much right from the standards. So what would it mean, in terms of talking, what would you want people to say and describe about what zeroes are in relationship to the polynomial functions?

ABIGAIL: I guess it's hard for me to articulate it. Because that idea coming straight from the standards is still very [procedural]. I don't know, it's like, "Well, we found real zeroes, and it means that we're gonna plot them, and then we're gonna talk about end behavior, and then we're gonna – Like, it's hard for me to come up with questions, honestly.

More brainstorming ensued, with Lani offering a way to emphasize the meaning of polynomials' roots:

LANI: I wonder – I mean this is, like, completely sacrificial – I wonder if emphasizing the meanings of "factor of zero" would help a little bit in terms of making the conversations less procedural. It's a totally untestable conjecture, so I have no idea. It might still be the same. But, like I don't know.

NADAV: And maybe, even the imaginary, you know? So we can't see them in the graph, but if we plug them into the equation, they will get –

ABIGAIL: It makes a zero.

LANI: It still makes a zero. Right.

NADAV: It makes a zero. So you can't say then, that i is not a zero.

Lani and Nadav proposed that the role of factors in yielding zeroes in polynomial equations might help clarify why imaginary roots are, in fact, roots. This raised a mathematical question for Abigail, another one we had anticipated in our planning:

ABIGAIL: Is it meaningful in terms of the graph at all? Is there a way that I can connect that?

LANI: No, we asked [the PDO's mathematician]. We asked [the mathematician]. [*laughter*]

NADAV: No.

LANI: This is part of our rabbit hole. We're like, "Hey, [mathematician]! We don't remember, but is there?" He's like, "No. There's not. You can't see it in a Cartesian plane." So.

ABIGAIL: Right.

BRAD: But some of them, I mean like the one they have with the double root – just the straight parabola – The other ones will do some funky, like, like this. [*Brad gestures with his hands*]

In this exchange, Abigail considered what it means to find roots with the graphical strategy that had been proposed earlier. Since the graphs of polynomials in the Cartesian plane do not have straightforward connections to their imaginary roots, Abigail was left with unresolved questions about how to proceed, albeit with new understandings of how students' mathematical talk was supported in her lesson.

Summarizing the Evolving VFF Questions

During this VFF debrief, the question of supporting students' conceptual talk was revised and specified as the teachers explored the video clips with the research team. First, Abigail conjectured that students' talk remained procedural because of her pacing. As she talked through this idea, she recognized that time did not guarantee students would engage in greater exploration, so perhaps she could have prompted them to explain why their solutions worked. Lani pressed to consider the mathematics of the task, which raised additional issues. Abigail even suggested that she was unsure about what was worth discussing about zeroes of quartics – especially imaginary zeroes, which have no obvious graphical representation. This shifted the conceptualization of supporting student talk as not only an issue of facilitation – through conversational routines or time allocation – but also an issue of the mathematics itself. Lani's last question pressed the teachers to consider what they wanted to hear students say about zeroes of polynomials, a question that the teachers did not have an immediate response to. The general concern about students' conceptual talk was refined to specific, potentially actionable issues. If teachers gain clarity on what they want students to understand about the topic, they can design activities – including mathematical tasks, discussion prompts, and discourse structures – to better support that understanding.

Later, Abigail said that something she appreciated about the VFFs was that she could reflect on how to provide "opportunities for students to interact with [math] content in a deeper way" (*Exit Interview, May 2019*). She said that, before Noether VFF 2, she had been wondering about the purpose of some topics that she taught; the VFF allowed her to reflect on existential questions about mathematical topics and to bring her instructional practices into better alignment with her goals. By reflecting on her pedagogical responsibilities in Algebra 2, and with rich evidence of how students interacted with her pedagogical actions, Abigail refined her pedagogical judgment and reconceptualized what she really wanted in students' conceptual talk.

What did SIGMa Teachers Talk About – and Did it Support Their Learning?

We delved into Abigail's VFF to illustrate how VFFs supported teachers' reconceptualization of core aspects of instruction. Zooming out to the entire data corpus of VFF debriefs, we now describe the kinds of co-inquiry topics these conversations supported. After each debrief, we coded[7] transcripts of the conversations for (1) the problems of practice that were discussed and (2) who raised them. Some of the problems of practice were introduced before the VFFs, like Abigail's initial question of, "What are students talking about during partner time?" Others emerged during the debrief, as questions or extended descriptions of dilemmas. Since problem framing can illuminate teachers' concepts, we used this coding as a "heat map" to identify which problems were raised and by whom. We could then show how teachers raised problems over time; how their framings shifted; and what kinds of problems of practice – or PoPs, as we called them – the VFFs supported.

In this section, we describe how we recognized PoPs in VFF debriefs. Using the summary of these codes, we show that VFFs supported teachers' learning opportunities around interactive dimensions of practice.

Identifying Teachers' Problems of Practice

As researchers, we wondered how much our facilitation made space for teachers' ideas and interpretations. We knew that we interjected new problem framings, as when Lani raised Emergent Debrief Questions 3 and 5 with Abigail. However, a serious concern would arise about our design conjectures and our guiding principles for facilitation if we were oversteering the conversations. To investigate the extent to which our research team raised dilemmas of teaching compared to teacher participants, we analyzed the PoPs that surfaced in VFF debriefs. Drawing on earlier work, we identified PoPs "through linguistic and paralinguistic cues that signaled classroom interactions experienced as troublesome, challenging, confusing, recurrent, unexpectedly interesting, or otherwise worthy of comment" (Horn & Little, 2010, p. 189). Because the PoPs often went beyond classroom interactions, we expanded our coding to reflect this broader terrain of inquiry, including topics such as planning, course design, and constraints of contexts, as well as dilemmas about the moral, ethical, and relational dimensions of teaching.[8]

What Teachers Learned About

Across 33[9] debriefs, we identified more than 3500 PoPs ($n = 3793$; utterances were multiply-coded if they addressed multiple PoPs). We categorized the PoPs into broad categories, summarized in Table 5.1. In the following

TABLE 5.1 PoPs coding summary in descending order of frequency.

Code Category	Instances of the Category	VFFs in Which it was Mentioned
Facilitation	1096 (29%)	33 (100%)
Planning and Curriculum	855 (22%)	33 (100%)
Facilitating Groupwork	583 (15%)	32 (97%)
Math-Specific	533 (14%)	32 (97%)
Social-Emotional	243 (6%)	29 (87%)
Demands of Teaching	190 (5%)	30 (90%)
Social	149 (4%)	26 (79%)
Assessment	100 (3%)	27 (82%)
Ethical	30 (1%)	13 (39%)
Physical and Material Constraints	14 (<1%)	9 (27%)

Note: "Instances" refers to the number of times each PoP category was brought up, across all VFFs. We also report the number of VFFs that included each category to show the distribution of PoPs across debriefs.

sections, we describe the issues captured in each code and the frequency with which they arose.

Facilitation

The most common PoPs addressed facilitation. This was a large category, including 35 codes that comprised 29% of all PoPs. Notably, PoPs about group-work facilitation are a separate category; Facilitation PoPs addressed questions about fostering student learning outside of groupwork (e.g., in whole-class discussions). This included very specific issues (e.g., *How do I avoid leading questions?* or, *How do I wrap up a lesson?*) as well as more general ones. The most common Facilitation codes were: *How do I foster student discourse? How do I pace a task well?* and *How do I make instructional decisions?*

The frequency and nature of facilitation questions suggest two things about our participants and the VFFs. First, given the intensity of our participants' involvement in professional development activities, these facilitation questions – which are about context-sensitive and interactive dimensions of practice – are ones that are not adequately addressed (and maybe not completely addressable) in traditional professional development. Second, the rich representations of practice may have uncovered and supported discussions about facilitation issues, making them more salient to teachers.

Planning and Curriculum

The second most common PoP category engaged issues of planning and curriculum (22%). This category included 18 codes about teachers' planning prior to instruction. The most common codes in this category were: *How do I design tasks? How do I align instruction to learning goals?* and *How can I plan so that content is accessible for all students?*

Since VFF debriefs focused on active instruction, it is notable that planning and curriculum came up so frequently; all coded debriefs included PoPs in this category. Of course, planning and curriculum – the substrate on which instruction is enacted – was invariably implicated in teachers' sensemaking.

Facilitating Groupwork

As we mentioned, we separated facilitating groupwork from general facilitation, simply because it came up so frequently in its own right. This category included 12 codes about facilitating groupwork to support student learning, ranging from logistical concerns (e.g., seating arrangements) to broader practices (e.g., fostering interdependence). The most common codes were: *How do I foster equitable participation? How do I approach groups in conversation?* and *How do I set up effective groupwork?*

Like general Facilitation PoPs, the frequency of Facilitating Groupwork PoPs suggests that this topic may be inadequately addressed in traditional professional development. Moreover, the VFF process likely uncovered details about facilitating groupwork that teachers wanted to investigate.

Math-Specific

The next most common category addressed specific mathematics content (14%), including the questions about factoring quartics in Noether VFF 2. This category included nine codes about learning specific mathematical practices or topics. The most common codes were: *How do I foster conceptual understanding? How do I help students think conceptually rather than procedurally?* and *How do I manage students' misunderstandings?*

In interviews, participants remarked on the value of our teams' experience in math education; we enjoyed diving into the details of content with teachers. The relatively high frequency of Math-Specific PoPs may have reflected teachers' eagerness to carefully consider this dimension of their instruction with colleagues and with us.

Social-Emotional

Social-emotional issues comprised 6% of PoPs. This category included seven codes about student's social needs, both inside and outside the classroom. The most common codes were: *How do I consider particular social dynamics?* and *How do I support students' confidence?* Although social-emotional issues were only 6% of the PoPs, this category arose in 87% of VFFs. The high frequency of these PoPs suggests that the video may have captured students' interactions in ways that made these issues salient for teachers.

Demands of Teaching

About 5% of PoPs addressed the demands of teaching. This category included five codes about professional requirements that were often outside teachers' control, like limited planning time. These issues arose in 90% of VFFs. This suggests that the demands of teaching were ever-present in, but not the focus of, debriefs. The largest code, representing 81% of PoPs in this category, was *How do I manage the constraints and demands of teaching?* Teachers' sensemaking almost inevitably invoked the institutional constraints on their practice.

Social, Assessment, Ethical, and Physical and Material Constraints

The last four PoPs categories were Social, Assessment, Ethical, and Physical and Material Constraints. Each of these topics constituted less than

5% of PoPs, but they varied in their distribution across VFFs, with Social and Assessment coming up in most debriefs and Ethical, and Physical and Material Constraints coming up less than half the time.

The Social category included 10 codes about social goals of teaching and relationships with and among students. The most common code − *How do I navigate status issues?* − represented 73% of PoPs in this category. The Assessment category included three codes about formatively and summatively assessing students. The Ethical category included three codes about teachers' moral and ethical dilemmas. These questions were often philosophical in nature, addressing issues like, *How much time do I spend preparing students for a test I don't endorse?* and *How do I structure student collaboration while respecting those who need more processing time?* Finally, the Physical and Material Constraints category included two codes about managing classroom layouts and limited material resources.

Trends Across PoPs Categories

The topics raised during VFF debriefs offer a heat map of the issues that became salient through our co-inquiry process. But an important question remains: Who was turning up the heat? Since our coding also accounted for who initiated any given PoP − focal teachers, partner teachers, or researchers − we identified a few important trends. First, we met our goal to center teachers' queries, from both focal teachers and their partners. Our data show that teachers raised twice as many PoPs as did researchers in the two highest-occurring categories (Facilitation and Planning and Curriculum). Among Groupwork and Math-Specific PoPs, teachers raised PoPs at a 3:2 ratio compared to researchers. In all but a few categories, focal teachers raised more PoPs than researchers.

A second important pattern was the consistency of the kinds of PoPs raised by focal teachers compared to those raised by researchers. In Planning and Curriculum, which accounted for almost one-third of PoPs, teachers and researchers consistently asked similar questions. Both groups focused on task design, questioning strategies, and resources to support students' sensemaking. Overall, these questions focused VFF debriefs on helping students make sense of mathematics and aligning instruction to student thinking, two important goals for ambitious and equitable mathematics instruction.

Within the Math-Specific category, teachers and researchers again raised similar questions, often about fostering students' understanding, prioritizing students' conceptual thinking (as in Noether VFF 2), and helping students develop mathematical connections − both between mathematical ideas and across parts of the task. This trend in PoPs supports our earlier claim that participants' commitments aligned with the goals of ambitious and equitable mathematics instruction.

In Facilitating Groupwork, the teachers' and researchers' questions diverged. Teachers typically raised questions about fostering equitable participation within groups, orienting students toward each other as intellectual resources, and deciding how to approach groups. In contrast, researchers raised issues about fostering student interdependence somewhat more often than teachers (26 instances from researchers, versus 22 instances from teachers), perhaps because it was an idea that we were more familiar with as a groupwork strategy.

In the Social category, the most prevalent PoP was about addressing status issues among students. This suggests that SIGMa teachers were highly attuned to the importance of academic and social status in their classrooms and that they sought to support students' equitable participation in mathematics.

PoPs Supported Ambitious and Equitable Mathematics Teaching

What do the PoPs tell us about our participants and their learning? We take teachers' questions as evidence of their familiarity with the cultural model of ambitious and equitable mathematics instruction. Overwhelmingly, our teachers brought up PoPs that were related to the shifts we outlined in Chapter 2.

This is not entirely surprising: As members of the PDO, they were part of a community of teachers committed to ambitious and equitable instruction. In Chapter 4, we noted that PDO teachers often attended – and led – sessions about teaching toward conceptual understanding, facilitating classroom discourse, cultivating inclusive learning environments, and so on. As Munter and Correnti (2017) demonstrated, teachers' instructional visions – the aspirations and ideals they have for teaching – are positively related to their instructional improvement. In other words, the fact that SIGMa teachers had ways of describing ambitious and equitable mathematics instruction suggests that they were ready to strengthen its enactment in their classrooms.

PoPs Addressed Interactional Aspects of Ambitious and Equitable Mathematics Teaching

With that framing in mind, we note that the most common PoPs – both those that occurred most often and in the most VFF debriefs – addressed the interactional aspects of ambitious and equitable mathematics teaching. SIGMa teachers came to the project with well-formed ideas about teaching; their early inquiry questions described goals that aligned with ambitious and equitable teaching, like Abigail's question about supporting students' understanding. Nonetheless, they had questions about classroom interactions related to those goals, like: *How should I respond to students' unexpected approaches? How can I better facilitate equitable groupwork?* The questions that they raised – and that the VFF design helped them consider – were rooted in the particulars of their classroom communities and teaching practices.

Discussion

In this chapter, we described how we put VFFs into practice. In producing rich representations of practice for VFFs, we created new tensions. The video and audio recordings offered substantive resources to anchor co-inquiry discussions, yet in creating these durable records of what are typically evanescent moments, we increased the social risk to participating teachers. For our work together to proceed effectively, we needed to foster trust, in part by assigning a lead researcher to each teacher team. This design brought up new dilemmas: We needed to ensure that VFFs bore some family resemblance to one another to study them as a context for teacher learning. For that reason, we coordinated our facilitation strategies by doing team reviews and building on prior research.

We then illustrated our facilitation by sharing Abigail's evolving questions in Noether VFF 2 (summarized in Table 5.2). Abigail's initial questions were about the quality of student talk about factoring quartic functions. From the video, Abigail realized that students' talk remained fairly procedural throughout the activity. With this in mind, her inquiry became increasingly specific. Abigail first wondered if students' procedural talk was a result of limited time, linking her pacing (pedagogical action) to the quality of their engagement

TABLE 5.2 The evolution of questions in Noether VFF 2.

Question	Speaker	Notes
What are students talking about during partner time? How can I better facilitate student-student talk so that they have a chance to explore ideas verbally? How can I better facilitate how students explain procedures?	Abigail	These questions are general instructional questions, not tethered to a specific mathematical topic other than "procedures."
Would students benefit from more time?	Abigail	After watching student talk stay fairly procedural, Abigail wonders if giving them more time to allow for more conceptual talk.
What do students understand about what a zero is?	Lani	Eliciting teachers' sense of students' current understandings of the core concept in the lesson.
What is the benefit of discussion imaginary zeroes?	Abigail	As the group contemplates students' understandings of zeroes, Abigail wonders about how discussing imaginary zeroes would support their learning.
What should students say and understand about zeroes?	Lani	Seeking clarification about what conceptual talk would look like for these problems by eliciting an image of students' understanding.

(pedagogical responsibility). With this question, Abigail noted the misalignment between her actions and her goals; she reasoned about its source but was dissatisfied in the diagnosis. After further reflection and video review, Lani pressed the group to consider the mathematics, further refining the collective inquiry by inviting teachers to describe their students' understanding of zeroes of polynomials. By drawing out representations of their desired practice – what would students' understandings even sound like? – the group had more interpretive resources for sensemaking. The group recognized the complexity[10] of describing imaginary zeroes, which have a clear algebraic meaning but are less obvious graphically. Abigail then asked about the broader purpose of the lesson, inviting a reflection on her and Greg's pedagogical responsibility to teach the topic. Lani tied that reflection to the role of this lesson in the larger curriculum: Where does this lesson fit into students' broader understanding about the zeroes of polynomials? This invited the teachers to consider larger timescales in weighing their pedagogical responsibilities.

VFF debriefs typically followed this pattern, with the focal teacher's inquiry question growing more specific as the group collectively interpreted the video. Debriefs usually included some combination of filling in representations of practice with additional details, brainstorming to elicit alternative pedagogical actions, and describing similar teaching experiences.

To better understand the terrain that VFFs covered, we zoomed out to the broader data corpus and shared the PoPs participants raised during VFFs, including both initial co-inquiry questions and topics that emerged during debriefs. We claim that the debriefs surfaced crucial issues about ambitious and equitable mathematics instruction and that teachers' engagement suggested they were asking questions about important aspects of their work. Most relevant for teachers' concept development, the VFF debrief dialogue offered teachers the chance to reflect on their pedagogical actions alongside their pedagogical responsibilities. This pedagogical reasoning, rooted in rich representations of practice and drawing on the collective interpretive resources of their colleagues and the facilitators, gave teachers opportunities to deepen their understanding of the relationships among different facets of instruction, thereby supporting the integrated knowledge that is a hallmark of responsive practice. Since our participants already showed high levels of commitment to ambitious and equitable teaching, we argue that the learning we document in the next three chapters offers evidence of how challenging an endeavor this is.

Notes

1 Recently, scholars have introduced models for in-the-moment instructional coaching as notable exceptions. For examples, see the work of Gibbons and colleagues (2017) on "teacher time outs" and Munson and Dyer's (2020) work on side-by-side coaching.

2 Abigail referenced our goal to share a version of this process with the PDO to support their work after our research ended.
3 In our transcription and narration of VFFs, we use the names they called us. Ilana Horn went by Lani, Patricia Buenrostro went by Patty, and Samantha Marshall went by Sammie.
4 All student names are pseudonyms.
5 Lani was referring to the comparatively short class periods at Noether, which were less than an hour, making inquiry teaching difficult as the time frame limited exploration.
6 An imaginary zero (or root) is a solution to an algebraic equation that includes the square root of a negative number. i is used to denote the square root of negative one. While these are solutions to the equation, they do not have the same graphical representation as real (i.e., non-imaginary) zeroes.
7 This analysis team was led by Samantha Marshall and Patricia Buenrostro, with contributions from Jessica Moses, Mariah Harmon, Natalie Boyd, Alec Macartney, and Ilana Horn.
8 PoPs were identified as segments of talk by one or more speakers. Some PoPs were established in one turn (e.g., "How could I have asked a better question?"); others developed dialogically, over multiple turns and speakers. For consistency, transcripts were double-coded.
9 Two debriefs could not be coded because of difficulties with the video record.
10 Pun intended.

References

Bannister, N. A. (2015). Reframing practice: Teacher learning through interactions in a collaborative group. *Journal of the Learning Sciences, 24*(3), 347–372.

Hall, R., & Horn, I. S. (2012). Talk and conceptual change at work: Adequate representation and epistemic stance in a comparative analysis of statistical consulting and teacher workgroups. *Mind, Culture, and Activity, 19*(3), 240–258.

Horn, I. S., & Campbell, S. S. (2015). Developing pedagogical judgment in novice teachers: Mediated field experience as a pedagogy for teacher education. *Pedagogies: An International Journal, 10*(2), 149–176.

Horn, I. S., & Kane, B. D. (2015). Opportunities for professional learning in mathematics teacher workgroup conversations: Relationships to instructional expertise. *Journal of the Learning Sciences, 24*(3), 373–418.

Horn, I. S., & Little, J. W. (2010). Attending to problems of practice: Routines and resources for professional learning in teachers' workplace interactions. *American Educational Research Journal, 47*(1), 181–217.

Horn, I. S., Garner, B., Kane, B. D., & Brasel, J. (2017). A taxonomy of instructional learning opportunities in teachers' workgroup conversations. *Journal of Teacher Education, 68*(1), 41–54.

Jackson, K., Gibbons, L., & Sharpe, C. (2017). Teachers' views of students' mathematical capabilities: Challenges and possibilities for ambitious reform. *Teachers College Record, 119*(7), 1–43.

Jurow, S., Horn, I. S., & Philip, T. M. (2019). Re-mediating knowledge infrastructures: A site for innovation in teacher education. *Journal of Education for Teaching, 45*(1), 82–96.

Little, J. W. (2003). Inside teacher community: Representations of classroom practice. *Teachers College Record, 105*(6), 913–945.

Munson, J. & Dyer, E. (2020). Collaborative sensemaking through side-by-side coaching: Examining in-the-moment discursive reasoning opportunities for teachers and coaches. In Gresalfi, M. & Horn, I. S. (Eds.), *The Interdisciplinarity of the Learning Sciences, 14th International Conference of the Learning Sciences (ICLS) 2020* (Vol. 4, pp. 1831–1838). Nashville, Tennessee: International Society of the Learning Sciences.

Munter, C., & Correnti, R. (2017). Examining relations between mathematics teachers' instructional vision and knowledge and change in practice. *American Journal of Education, 123*(2), 171–202.

Yin, R. K. (2012). Case study methods. In H. Cooper, P. M. Camic, D. L. Long, A. T. Panter, D. Rindskopf, & K. J. Sher (Eds.), *APA handbook of research methods in psychology, Vol. 2. Research designs: Quantitative, qualitative, neuropsychological, and biological* (pp. 141–155). American Psychological Association.

Zeichner, K. (2010). Rethinking the connections between campus courses and field experiences in college-and university-based teacher education. *Journal of Teacher Education, 61*(1–2), 89–99.

6
LEARNING TO INQUIRE INTO TEACHING

With Grace A. Chen and Katherine Schneeberger McGugan

In this chapter, we describe how Project SIGMa teachers learned how to learn in the video formative feedback (VFF) process. As we described in Chapters 4 and 5, VFF debriefs aimed for exploratory dialogue about teaching, which is not typical in professional development experiences. As a result, our participating teachers needed to develop new ways to engage with and learn from VFFs. To illustrate this, we focus on three teachers' learning to learn: Greg Kahae, Lizette McLoughlin, and Veronica Kennedy.

Our account of their learning to learn shows the varied paths teachers traversed as they developed new forms of learning – their various learning trajectories – suggesting that similar shifts might be required of other teachers engaging in collaborative sensemaking as professional development (CSPD; Ehrenfeld, 2021). Specifically, we describe how these teachers renegotiated subject-world relations as they modified their understandings of what teaching entails and how people continue to learn about it, opening possibilities for new modes of learning about practice.

Teachers' Learning to Learn with Video: Moving from Evaluation to Interpretation

While our co-inquiry stance (described in Chapter 4) may distinguish our version of CSPD from other examples of video-based teacher learning activities, our findings nonetheless confirm a prominent theme in prior research. Others have noted that, as teachers grow more comfortable with inquiry-into-practice through classroom video, they shift from evaluative to interpretive frames (Luna & Sherin, 2017; van Es, 2012), moving away from binary considerations of good/bad teaching toward an analysis of practice (Borko et al.,

DOI: 10.4324/9781003182214-8

2006). Other researchers have fostered such environments through video-viewing activities that emphasize inquiry (Borko et al., 2015) and by otherwise communicating the importance of teachers' learning over evaluation (Zhang et al., 2011). Through video review, teachers' conversations become more dialogic (Coles, 2013), allowing them to focus on important issues, like students' mathematical thinking (van Es & Sherin, 2010), its relationship to curriculum (van Es, 2012), and how to meet the needs of different learners (Steeg, 2016).

In our framework, this shift from evaluative to exploratory talk is itself a reconceptualization of what it means to discuss teaching. As teachers move away from sharing tips and tricks to having more analytic, conjecture-laden conversations, they create opportunities for concept development about instruction (Horn et al., 2017). In a comparison of teacher workgroup discussions, Ilana Horn and Britnie Kane (2015) found that more accomplished groups discussed teaching in exploratory ways; this gave them more opportunities for learning as compared to less-accomplished groups. Stated plainly, teachers who had rich ways of talking about instruction gained more from talking about instruction, a classic accumulated-advantage phenomenon.

In theoretical terms, teachers with onto-epistemic views on teaching as interpretive and uncertain are better prepared to learn with colleagues about teaching, since their interpretive stance makes teaching choices discussable. This discussability, in turn, is crucial to opening dialogue for teachers' concept development about pedagogical actions. In this chapter, we describe how participating in VFFs supported teachers' shifts toward an interpretive stance on teaching and how that shift, in turn, supported their learning.

The Baggage of the Binary: Renegotiating Subject-World Relations

As research on teacher learning shows, evaluative stances are common among teachers. From a learning perspective, evaluation often reflects a particular subject-world relation, one in which pedagogical actions can be confidently judged as either good or bad, despite limited evidence. As narratives of Greg, Lizette, and Veronica will show, subject-world renegotiation often invokes questions about what it means to be a good teacher – contested, varied, and situational notions that are encapsulated by Competing Visions of Quality (Chapter 1). Specifically, teachers confronted what we came to call the *baggage of the binary* – the notion that the "goodness" of a teaching practice is stable, superseding contextual particulars and existing in opposition to (and in mutual exclusion of) its "badness." In contrast, shifting toward an interpretive stance invited examinations of the consequences of specific teaching practices, asking: What goals does this practice support? What goals are not supported?

Many societal messages support the onto-epistemics of the binary, making it common-sense. Qualitatively, the language of "best practices" for instruction (e.g., Lemov, 2014) implies that some practices are more "good" than others, regardless of grade level, content area, culture, or other contexts. Quantitatively, value-added measures of teaching rank and sort teachers as inputs and (narrow versions of) student learning as outputs, adding a technocratic certainty about goodness in teaching (Biesta, 2015). These measures create narratives about teachers' verifiable goodness, occasionally leading to tragic outcomes for individual teachers when rankings are made public (Casey, 2012; Lovett, 2010).

The good/bad binary is a socially, institutionally, and historically embedded concept. In its ubiquity, it reflects the zero-point epistemology about knowledge in teaching (see Chapter 4) – the neo-positivistic idea that we can nail down good teaching and disseminate best practices without regard to the particulars of people, communities, and contexts (Biesta, 2007). This epistemology becomes a keystone in teachers' conceptual systems, and it works against onto-epistemic heterogeneity – the notion that there are multiple, reasonable perspectives worth engaging – which creates the dialogue that is critical for teachers' concept development. Of course, most teachers accept some variation in pedagogical approaches, often captured in the folk wisdom of "teaching styles." However, the teaching styles construct still does not typically invite teachers with different perspectives to engage in dialogue. Rather, in teachers' conversations, a mention of teaching styles usually forecloses dialogue; it is a call to "live and let live," invoking norms of non-interference among teachers (Little, 1990).

Using dichotomous charts (Haraway, 1988), we illustrate the onto-epistemic reconceptualizations that are involved in shifting from the good/bad binary toward expansive views about teaching (Table 6.1). As in Chapter 2, this dichotomous chart represents two poles, highlighting the tensions teachers must navigate: On the left is a neo-positivistic, zero-point epistemology that affirms the binary perspective of teaching as good or bad. On the right is an expansive view of good teaching as irreducibly situated and contingent.

The shift from evaluative to interpretive talk about teaching is a crucial gateway to transformational professional learning. In contrast to models of professional development that emphasize teachers' mastery or mimicry of specific teaching practices or routines, reviewing video can provide opportunities for teachers to consider and reconcile contradictory ideas about different aspects of teaching and weigh them alongside their various (sometimes competing) goals for instruction. Teachers can develop pedagogical concepts as they contemplate the consequences of pedagogical actions, generate alternatives for working differently, and consider how these might reflect (or conflict with) their values. However, the binary view of teaching – where practices are good or bad because researchers, colleagues, or other authorities said so – makes it harder for teachers to exercise reasoned judgment and make thoughtful comparisons among possible courses of action.

TABLE 6.1 Core onto-epistemic tensions in conceptions of good teaching.

Binary View of Good Teaching	Expansive View of Good Teaching
Good teaching can be broken down into specific practices, curriculum, and activities that will result in student learning.	Research suggests that some practices, curricula, and routines more effectively support student learning than others. However, even research-backed pedagogies have limitations, and teachers must exercise professional judgment to listen, respond, and adapt these pedagogies for their students.
Good teaching is the result of an individual, accomplished teachers' skill and knowledge	Good teaching is relational, resulting from a mix of teachers' skills, knowledge, and their capacity to thoughtfully respond to their students, who may be more or less receptive to learning with them.
Good teachers always have good lessons.	Lessons are an interactional accomplishment, with teachers designing experiences based on what they know of their particular students, classroom dynamics, content, and sociocultural settings in relation to what they aim for students to understand.
Good lesson plans lead to smooth lessons.	Good lessons are thoughtfully designed, with formative assessment at key moments shaping their emergent enactment as teachers listen and respond to their students.

Undoubtedly, our perspective builds on an assumption of teaching's irreducible situativity, thus embracing onto-epistemic heterogeneity: While some forms of teaching may be generally better or worse, "good" teaching does not always look the same with every teacher, in every classroom, with every group of students, and in every sociocultural setting. This is not a relativistic position that effectively renders all research useless – we are not asserting that all teaching is equally good. Rather, we insist that any claims of goodness can only be verified by specifying whom that goodness serves and toward what ends. In other words, "good teaching" has many meanings. As Gary Fenstermacher and Virginia Richardson (2005) elaborate, "successful" teaching – meaning students learn the intended content – may be different from "quality" teaching. A teacher can "succeed" through bullying, threats, and other forms of harm, but we would not deem those methods to be worth emulating. Because teaching, as an activity, serves many simultaneous goals and has many simultaneous consequences, it is common for some goals to be met while others are neglected –– or even undermined. As a result, judgments of quality are complex and dependent on the values and priorities of the person making them. For instance, in our work, we take the stance that practices supporting more equitable student outcomes are "better" than practices that do not, but we are careful not to assert that this means such practices look the same in all settings or unequivocally serve all possible teaching goals equally well.

An expansive view of good teaching points to the extent to which teachers' commitments, epistemologies, and habits of mind supersede many details about what they *do*. That is, the "goodness" of teaching is ultimately relational, a judgment to be made with careful consideration of particular teachers, learners, places, and historical moments. Reflecting the Action Premise – the subject-world relation that values teachers' actions over their understandings – conversations about teaching tend to treat what teachers *do* as the most important aspect of instruction, making deeper explorations about their reasoned judgments rare (Horn et al., 2017). What keeps relativism at bay in reflective teachers' conversations is serious engagement with pedagogical responsibility – weighing the institutional, legal, and ethical consequences of actions' meanings (Chen et al., 2021).

Many messages about good teaching circulate in teachers' workplaces. An incomplete catalog of these messages includes: *Good teachers produce good student test scores. Good teachers have good classroom management. Good teachers stay on pace with the curriculum. Good teachers follow students' interests. Good teachers nurture students' humanity. Good teachers are highly knowledgeable. Good teachers understand students' development of disciplinary ideas. Good teachers affirm students' racial, cultural, linguistic, and gender identities.* While these messages are not all mutually exclusive, they tap into different onto-epistemics of teaching and learning (Cohen, 2011), conveying different emphases for teachers' work and, consequently, for their learning.

Notably, these messages do not exist on a menu from which teachers select based on personal preferences – or their "teaching styles" – because not all messages about good teaching have the same social power. In the U.S., recent neoliberal educational policies – like merit pay and "value-added" scores – are rooted in the assumption that teachers' goodness/badness is the most important lever for educational change (Kumashiro, 2012), despite mounting evidence that school cultures, discipline policies, course placement, supplemental education resources, school funding, families' economic precarity, and myriad other factors shape students' experiences in school. Nonetheless, teacher-centered neoliberal policies have tremendous influence due to their material consequences: Teachers can lose their jobs or have their schools reconstituted if they do not meet the expected outcomes. These consequences infuse some messages (*good teachers produce good test scores*) with greater salience than others (*good teachers follow students' interests*). When the logics of neoliberal policies permeate the conceptual infrastructure of schooling and teachers' work, the notion of individual teachers as ultimately responsible for educational outcomes becomes a commonsense, hegemonic frame for understanding teaching (Horn, 2018) – a prevalent, institutionally-endorsed onto-epistemic stance – adding further psychological weight to the baggage of the binary. Meanwhile, other potentially informative outcomes that would implicate broader systems, such as graduation rates, college preparatory courses, students' social-emotional well-being, student learning in contexts other than testing, and so on – receive much less attention from policymakers and the public.

Moreover, notions of good teaching cut close to teachers' identities – how they understand themselves as teachers. As we described in Chapter 3, teachers' identities are deeply related to their pedagogical responsibilities, the commitments that both guide and anchor their pedagogical reasoning (Chen et al., 2018). Teachers' identities reflect both individual proclivities and narratives about teaching that come from schools, professional groups, and society writ large (Horn et al., 2008). When teachers discover that they are not meeting their core commitments, this can threaten their very identities as teachers, as we narrate in Lizette's case.

If the zero-point epistemology about teaching were true, then it would follow that teachers are, in fact, rankably better or worse, justifying anxieties about their demonstrable goodness or badness. Indeed, the baggage of the binary generated by this logic is burdensome, posing potential face-threats to teachers inquiring into their practice (Vedder-Weiss et al., 2019), as they risk being "found out" as not as good as others presumed. In other words, the baggage of the binary fuels teachers' imposter syndrome. Because our participants were experienced teachers who were highly respected in their schools, the face-threat of the binary may have been especially keen – it would have been very damning for a veteran teacher of 20-something years to fall into the "bad" category. Thus, a big project of our co-inquiry approach was to help teachers resist the easy evaluations suggested by the binary and instead, make space to acknowledge that teaching is multidimensional; serves numerous, simultaneous (and often competing) goals; is enacted in relation to particular people, places, and moments; and its "goodness" exists in relation to these goals and contexts. Instead of taking specific moments as evidence of good (or bad) teaching, we prompted teachers to reason through the consequences of their pedagogical choices and whether those consequences aligned with their values.

Although the Competing Visions of Quality Premise was the most frequently renegotiated subject-world relation among Project SIGMa teachers as they learned to learn in the VFFs, we noted other shifts in their understandings of who they were in relation to their learning. For instance, by uncovering new facets of classroom life, VFFs frequently pushed back on teachers' impressions about how their lessons went. Recall that our team's video review often focused on students' interactions, especially those that took place outside teachers' immediate notice. Thus, VFF debriefs presented teachers with new forms of feedback about their instruction. This sometimes altered their sense of what productive instructional action entailed, leading to new understandings of which actions were useful (the Utility Premise). In other words, the close analysis of video sometimes uncovered details that smooth lessons masked (the Smoothness Corollary), such as when lessons went according to plan, but student conversation nonetheless reflected significant confusion about important

TABLE 6.2 Overview of Greg's, Lizette's, and Veronica's learning to learn in VFFs.

	Greg	Lizette	Veronica
Where they started	An experienced, traditional teacher who was protective of his colleagues	An avid teacher learner who embraced inquiry into practice	An enthusiastic, experimental teacher who changed schools and felt like a "freshman" again
How they learned to learn	Shifted from taking VFFs as evaluative to seeing them as an opportunity for co-inquiry and therefore for learning about his own practice	Recognized that difficult classroom moments represented just "part of" her teaching practice rather than indicting her as a bad teacher	Transitioned from framing teaching as epistemically uncertain to aleatorily uncertain
What helped them learn	• Colleagues' engagement and modeling • Trust in colleagues and PDO leaders • Making space for his identity	• Normalizing disappointing lessons • Providing additional time to respond	• Colleague characteristics • Time and comfort with VFFs • School support

ideas. As teachers renegotiated their subject-world relations, VFFs made space for them to reconceptualize their work. In the following sections, we share how Greg Kahae, Lizette McLoughlin, and Veronica Kennedy renegotiated the baggage of the binary through their VFFs.

Greg Kahae: Moving from Evaluation to Inquiry While Maintaining a Supportive Stance

Throughout our project, Greg Kahae learned two important things. First, he learned how to engage in collaborative inquiry into practice through VFFs, resolving some initial tensions he felt around supporting his colleagues while co-inquiring into their practice. As his participation in VFF activities deepened, he also learned something important about his own teaching – namely, to consider social dynamics in fostering students' small-group discourse. Here, we focus on Greg's learning about co-inquiry into practice, while describing how it created opportunities for him to learn about his own instruction.

Background and Context: An Experienced, Traditional Teacher Investigating Ambitious Teaching

At the time of our study, Greg Kahae had taught between 20 and 25 years. He had spent most of his career at Noether High School, primarily working in isolation. In that time, he had never experienced any kind of instructional coaching or substantive feedback from administrators. Several years before our study, Greg's isolation was broken when his junior colleagues, Brad Miller and Abigail Graham, encouraged him to join the PDO. During our study, Greg taught Geometry, AP Calculus, AP Computer Science Principles, and Algebra 2. As a veteran Noether teacher, he told us that students knew what to expect when they came to his class: "My reputation at Noether is, you better be ready, prepared to have a demanding teacher" (*Member Check Interview, March 2020*). Although he was proud of his long career, he held his junior colleagues, who came up in the PDO's model of ambitious and equitable mathematics instruction, in very high esteem. As he told us:

> I know that Brad and Abigail are excellent teachers. I know they're probably – I'll say this on record – probably like the best [PDO] teachers there are. They're so great. So, I know when I'm watching their videos, it's greatness.
>
> *(Exit Interview, May 2019)*

Sometimes, he even went so far as to compare himself to them unfavorably, saying that Brad and Abigail would not learn much from watching him teach. "I view myself as an average teacher," he said (*Exit Interview, May 2019*). He enjoyed learning with the PDO, explaining, "I've been an old dog. All I know is these old tricks. And now I go, 'Oooh!'" (*Exit Interview, May 2019*). By positioning himself as learning from his junior colleagues, Greg may have been speaking honestly, acting modestly, or seeking to mitigate unfavorable judgments about his teaching by not claiming "good teaching" as his own. In any case, it was a common refrain when he interacted with members of our research team, who repeatedly reassured him that we enjoyed spending time in his classroom, with his organized lessons, clear routines and expectations, and interesting mathematical tasks.

Where Greg's Learning Started: Reluctant Participation and Protectiveness of Colleagues

Although he was initially a reluctant participant, Greg was clearly committed to and supportive of his colleagues. Throughout Year 1, Greg participated as a partner teacher. At first, despite our team's efforts to frame the conversations as co-inquiry, he seemed to experience them as evaluative and defended his colleagues fiercely. For example, in Noether VFF 1, when Brad was the focal teacher, Greg's body language (arms crossed, leaning back) signaled mistrust and skepticism about the activity. His participation was minimal: He only spoke twice, both times offering positive comments about Brad's teaching (e.g., "I love how the kids are passionate!").

In Noether VFF 2, with Abigail as the focal teacher, Greg participated more (18 turns of talk). This was not entirely surprising, since he had co-planned the lesson with Abigail. The VFF debrief surfaced numerous issues around lesson design (see Chapter 5 for more details). For example, during the debrief, we asked about Abigail's goals for student learning. Greg chimed in several times to explain the rationale behind their planning, sometimes sounding defensive about how hard it was to plan an inquiry lesson around a dry topic (e.g., "it's hard to find these [*quartics that are factorable*], You know, you have to look through the other two books and find ones that work"). His comments seemed to defend the lesson not only on his behalf, but on Abigail's. Greg was a loyal colleague; he clearly championed Abigail as a teacher.

After the debrief, Lani checked in with Greg to reiterate the supportive intent of the VFF process. Greg said sardonically, "Thank you for overwhelming us with the 'Oh, what are we going to do?' factor." When she explained that this was not our intent, he amended, sounding conciliatory, "[The PDO] is always a little overwhelming." Later, in his Exit Interview, Greg recalled Noether VFF 2. He said that he and Abigail were "offended that [the lesson plan] was being questioned."[1] Although he seemed to experience Noether VFF 2 as evaluative, Greg continued to work with us.

Greg's Hybrid Practice of Supportive Interpretation

Greg's participation evolved by the end of Year 1, maintaining his supportive stance with colleagues but increasing his capacity to engage in co-inquiry. We saw this transition in Noether VFF 3, with Brad as the focal teacher. The quality of Greg's talk changed, as he exhibited a more exploratory stance toward the video clips. Specifically, Greg posed numerous questions to support the group's collective inquiry, such as ascertaining students' prior experiences (e.g., "Did they graph in vertex form yet?"), seeking interpretive grounding for understanding the class (e.g., "Is this the same period you videotaped last time?"), and suggesting questions Brad could ask his students (e.g., "Why do you think that?"). Greg's new mode of participation suggested that he took the debrief as an opportunity to reflect and brainstorm.

This pattern continued in Year 2, as Greg further developed his capacity to co-inquire into his colleagues' practice. In Noether VFF 4, the team watched Marisa's Geometry lesson on midsegments of a triangle. Marisa herself did not feel good about how the lesson went; as she wryly described it:

> You know how there are some days when you have a lesson that's planned and everything goes just right and perfect and the kids learn, they do what they're supposed to be doing, and your pacing is right on, everything just *goes*? (*Pause.*) This wasn't that lesson.
>
> (*Noether VFF 4 Debrief, December 2018*)

Together, we viewed a clip showing students talking about the task, and Marisa acknowledged that their discussion was better than she expected. Greg quickly jumped in to affirm her, noting that the activity and discussion were "bringing out things that they know about quadrilaterals. It's not a hundred percent correct, but things that are coming up, like parallel lines, what's a parallelogram?" In this interaction, he blended his inclination to support his colleagues with a grounded interpretation of the interaction we had viewed.

As we discussed the lesson, Marisa realized that she needed to reconsider her pacing: The introductory activity had taken more time than she anticipated, so she rushed through the rest of the lesson. Greg, who had taught Geometry many times, offered thoughtful advice for scaffolding her lesson to help students get to the crux of the activity. For instance, he proposed that she could support students' use of vocabulary terms like "alternate interior angles" by providing students with labeled diagrams to reference. In this way, Greg continued to support his colleagues while co-inquiring into their practice.

Going Focal: Greg Investigates His Own Teaching

By Noether VFF 5, Greg was ready to be a focal teacher; he invited us to film his AP Computer Science Principles class. He later confessed to feeling some trepidation in being filmed, deliberately asking our team to visit his AP class and not his Geometry class:

> You want this perfect classroom that's not… you want to show people that, okay, everything in your classroom is happy-go-lucky […] In fact [in Geometry], they're going to see a group here that's not going to be so great, so I'm afraid of that.
>
> *(Exit Interview, May 2019)*

In other words, Greg managed his discomfort at being filmed by inviting us to a lesson he felt confident in, a strategy that worked fine in our intervention design while keeping him on safer ground in relation to the binary.[2] During the lesson, students worked in pairs on a coding activity, with the goal of understanding the difference between local and global variables. Each pair shared a computer; one student was the "driver," meaning they controlled the keyboard, while the other was the "navigator," who directed the driver as they worked on the assignment. One part of the activity was particularly challenging. Students were given a buggy segment of code that included a function calling on the variable "count," which needed to be global for the function to run. However, in the code, "count" was initiated both globally and locally (within the function), which created an error. Students needed to debug the code by deleting the local initialization of "count."

For the VFF co-inquiry, Greg's question was about how pairs of students communicated. To investigate this during the debrief, we watched two clips of students collaborating. The first clip showed students talking about how to debug the code. As they talked, they correctly debugged the code, but without discussing the idea of local and global variables. Near the end of the clip, Greg checked in with them, asking, "Did you figure it out?" The pair said they had, and Greg left to check in with other students.

After watching the clip, one of Greg's colleagues, Joseph Park, noted a limitation of the task: Students could solve this activity without necessarily understanding the larger concept of local and global variables. Greg agreed and wondered aloud if he should have pressed for more explanation, suggesting, "I could have stayed. But it looked like they were working, so I was like 'Okay, I'll let them keep working.'" This illustrated a common tension in facilitating groupwork, as it is not always obvious when to interrupt groups and when to let them keep working.

The second clip showed Greg taking using a different approach to check on a group of three students – Dan, Leo, and Ethan. They had also debugged the code, but Greg pressed for more explanation, asking, "Why do you think that was messing it up?" Leo offered the first explanation, and then Greg asked Ethan to expand on what Leo said. Greg told Dan to take on the driver role but did not ask what he understood about the problem. Reflecting on the clip, Sammie noted that Greg had facilitated this interaction differently than the previous one, in that he had pressed students to explain their reasoning. But Greg pointed out that he only asked Leo and Ethan – and not Dan – to share. While Greg had tried to involve multiple students in the check-in, he did not have time to ask all three students about their sensemaking. As he reflected on this clip, he expressed a growing recognition of the complexity of facilitating students' collaborative discourse.

In his Exit Interview, Greg recalled that unlike Noether VFF 2 – when Abigail was the focal teacher and Greg felt offended about critiques of the lesson plan – he "felt good" about Noether VFF 5, noting that by watching video, he could see "those little details I don't know if I would be able to see without videotape, and seeing the interaction kind of opened my mind." (*Exit Interview, May 2019*). Through the VFF process, Greg identified questions to explore in his teaching and was able to investigate them using rich representations of practice and the conceptual resources afforded by his colleagues. Over the course of the project, Greg shifted from an evaluative stance to an interpretive co-inquiry stance. Reflecting on Noether VFF 5, he said, "I was, 'Oh, okay! the kids are talking with each other, and that's a good sign, and how can I improve the talking?'" (*Exit Interview, May 2019*).

He also told us that, because of the Noether VFF conversations, "I think a lot more about how I want to get the students to interact with each other" (*Exit Interview, May 2019*). He elaborated:

> Before doing any of this, and before working with Abigail, I never really thought about how the students interacted. [...] It was like, "Oh, they communicated. That's effective to me." But this whole process, it's a little overwhelming because, "Okay, I've got to deal with a little bit of a hierarchy with the kids."
>
> *(Exit Interview, May 2019)*

In other words, he recognized that his facilitation not only mattered for whether students talked to each other, but also for issues of "hierarchy" – or *status* (see Chapter 2) – that shape how students talk to each other. After learning how to learn through VFFs, Greg was willing to share his teaching with his colleagues and the research team, introducing new questions for his own practice. By the end of the project, Greg had shifted his classroom structures to give students more opportunities to think together, with greater attention to status issues.

What Supported Greg's Learning to Learn

Our account of Greg's learning to learn through VFFs illustrates how he renegotiated his subject-world relations, as well as the conditions that helped him do so. First, his participation shifted from reticent (Noether VFF 1) or defensive (Noether VFF 2) to involved in conversations about teaching (Noether VFFs 3-5). Especially in Noether VFF 2, as the group analyzed a lesson he had co-planned, his defensiveness seemed to reflect the baggage of the binary. In Noether VFF 5, his choice to examine a class where he felt confident – and not one that he was insecure about – suggested that he was still mindful of the binary, but he was comfortable enough with his colleagues and our team to share his practice.

Colleagues' Engagement and Modeling

Brad and Abigail, Greg's colleagues whom he greatly admired, modeled comfort with the VFFs, encouraging Greg's persistence in an initially uncomfortable experience. In fact, Brad and Abigail described enjoying the feedback from the VFFs. In Exit Interviews, they stood out in their comfort with the vulnerability of being recorded, welcoming the feedback and discussions about teaching supported by the VFF debriefs. Eventually, Greg learned to join these discussions. As we describe in Chapter 8, Brad's learning was evident as we watched it unfold, perhaps further legitimizing the VFF activity for Greg. As

Greg described, he came to see VFFs as a team endeavor: "We all go through that kind of vulnerability [of being filmed] together. I think it makes it so that everybody has their turn to be vulnerable" (*Exit Interview, May 2019*).

Trust in his Colleagues and the PDO Leaders

That Greg continued to work with our team – despite his initial reluctance – reflected his trust in Brad and Abigail, as well as in the PDO leaders. This was evidenced when he linked his "overwhelm" in Noether VFF 2 with a more general experience of learning in the PDO. Our partnership with PDO leaders likely gave us credibility that we may not have had otherwise. Greg explained:

> Before [the PDO] or before any collaboration, I was not really improving my teaching. I might've been improving the content of my own knowledge, but I wasn't improving as much as I could or as I am in the last few years with this collaboration
>
> *(Member Check Interview, March 2020)*

Although he sometimes found the process overwhelming or uncomfortable, Greg seemed to take satisfaction in learning and improving his teaching.

Making Space for Greg's Identity

Our co-inquiry stance and commitment to onto-epistemic heterogeneity made space for Greg's perspectives and experiences. Greg's initial responses to the VFFs reflected, in part, his loyalty and commitment to his colleagues. Unlike the evaluative stance on teaching, which we deliberately tried to move away from, we welcomed Greg's identity as a supportive and admiring colleague, and he found a productive place within the group's brainstorms. This ability to blend his supportive inclinations to the co-inquiry activity allowed Greg to integrate his teaching identity into our work together. In making this space, Greg negotiated his different commitments and found ways to learn.

Lizette McLoughlin: Managing the Emotional Baggage of the Binary

In many ways, Lizette McLoughlin was an ideal participant for Project SIGMa. She was, like all the PDO teachers, invested in her own learning. She described herself as "a problem-solver kind of person," suggesting she liked to jump in and figure things out, both in teaching and in mathematics. Lizette had a supportive partner teacher and quickly established strong rapport with

Brette, her primary Project SIGMa contact. Although she was clearly game for co-inquiry into practice, Lizette encountered unexpected vulnerabilities stemming from the binary.

Background and Context: An Avid Teacher Learner with a Strong School Partner

During our study, Lizette McLoughlin had taught for over a decade at Fermat High School. She was highly invested in her own professional learning and that of her colleagues; as a member of the school's instructional leadership team, Lizette designed and facilitated professional development for colleagues. She also had experience using video in professional development and found that experience valuable. Lizette was eager to reflect on and improve her teaching.

Lizette also had a strong relationship with her partner teacher, Julie Woodman. They had been friends and colleagues for many years and had developed similar pedagogical responsibilities. As Lizette described their commitments:

> We agree on a lot, in terms of what we think is important for our students. And so it's been really beneficial to talk some of those things through, because you're not arguing about it. But you also have someone who will check you and say, "Is that really what you think is important?"
>
> *(Interview 3, April 2018)*

While they usually taught different courses – Lizette in AP Calculus, Geometry, and Algebra, and Julie in Statistics and Algebra 2 – they often met to talk about teaching during their shared planning period. They encouraged and supported each other; this was apparent in VFF debriefs, as well. As Lizette described their VFF dynamic:

> It's special [...] We are such close friends, too, so I feel like a lot of times we were both sitting there [in VFF debriefs], like, trying to be each other's cheerleaders. Like, "You got this," and when the conversation would get kind of hard, being like, "Yeah, we still want to push this conversation but also want to make sure that, like, everything's good."
>
> *(Exit Interview, May 2019)*

The deep trust between Lizette and Julie allowed them to balance cheerleading and more critical reflection. This created a supportive environment for Lizette to investigate her teaching.

Where Lizette's Learning Started: Ready for Inquiry but Feeling Vulnerable

Although we did not know the extent of it when we began our work together, Lizette also felt vulnerable with the VFF process. She described the process of watching herself teach as "an emotional rollercoaster of, 'Here's a really good thing,' and you feel like, 'Oh my God, look at that cool thing I did!' And then, 'Here's the thing you're doing,' and it's like, 'Phew!' [*sounding disappointed*]" (*Exit Interview, May 2019*). Further elaborating her experience being the focal teacher, she said:

> It's halfway terrifying, halfway amazing. I remember just sweating, because I was so nervous and feeling so uncomfortable [...] You're very vulnerable, which is a really good thing, and also a really hard thing.
>
> (*Exit Interview, May 2019*)

In Lizette's vivid descriptions of her emotional experience in the VFFs, there is an echo of the baggage of the binary. She described certain practices as being "really good," and set up others as being bad (the disappointed "Phew!"). To some degree, Lizette's trust – both in Julie and in our research team – mitigated the vulnerability she felt but could not completely eliminate it.

Being Pulled Up Short: Learning to Manage the Baggage of the Binary

Lizette's vulnerability became highly visible during Fermat VFF 4, which was Lizette's third VFF as focal teacher. In this lesson, AP Calculus students worked in small groups on a task about position, velocity, and acceleration. The task and groupwork were typical of Lizette's teaching practice, but the class was unusually large. Instead of about 25 students working in seven groups, this class had 40 students working in 11 groups.

As we reviewed the video recording after the lesson, we recognized the consequences of the large class size. Even though Lizette spent most of the groupwork time working with students, the additional students – and additional groups – meant that Lizette had little time to spend with each group. Many groups had questions as they worked, so Lizette's check-ins with them were limited to brief one-on-one discussions with individual students. Despite her best efforts to answer everyone's questions, students in focal groups expressed frustration at having to wait for extended periods of time before getting Lizette's attention. Importantly, this was atypical from what we had seen of Lizette's practice. In previous VFFs, Lizette had involved the whole group when she checked in about a question. She had also spent time quietly circulating and listening to students as they worked.

As we prepared for the Fermat VFF 4 debrief, we attributed these changes to Lizette's unusually large class size. To us, it seemed like Lizette was facilitating groupwork as she typically did – letting students collaborate in groups, with little intervention or interruption – but that she had not adjusted her practices to account for the large class. This reflected our interpretive stance on teaching: Lizette's typical groupwork practices were not inherently bad, but they did not seem to serve her well in an overcrowded classroom.

During the debrief, we showed a video clip that challenged Lizette's core commitments as a teacher. In the clip, three students were interpreting a graph of a soccer ball's velocity over time. The students pointed to time intervals on the graph and features of the graph in order to determine the ball's position, speed, and acceleration. As the students discussed, they asked each other questions ("Is it positive?", "it's...slowing down?", "What is that?") but were not confident in their answers. One student, sounding somewhat frustrated, bluntly stated, "I'm just so confused."

After playing the clip, Brette described the group's frustration as typical of what we had observed in the class, saying that students were "working so hard, but they're spinning." Julie wondered if the group had asked for help. Brette said that they had raised their hands but got tired of waiting for Lizette to get to them; they eventually asked another group for help. Brette also pointed out the differences that we had seen in Lizette's groupwork monitoring – that she usually took time to involve the entire group in discussions and to listen to students' conversation before checking in with them. By raising these points, our goal was to illustrate how changes in the classroom setting – specifically, the increased number of students – tested the limits of Lizette's pedagogical actions. That is, we did not view Lizette's actions as bad; rather, they did not meet the needs of the large class.

But the video (and, likely, our summary of students "spinning" into unproductive exploration) cast a harsh light on the disconnect between Lizette's experience of the lesson and her students'. She observed that, "from a kid perspective, they're not getting very much help [...] but from my perspective, I'm constantly with someone." Because she was so active during the lesson – moving from group to group, answering student's questions – Lizette had not realized that her students had felt neglected.

Before the debrief, Lizette had reflected on the lesson and realized that it "didn't land," but she initially attributed students' difficulties to challenging content and a poorly written task. Importantly, the debrief supported this diagnosis and refined it: The content and task design likely contributed to students' frustrations, creating a situation where many groups needed Lizette's help. But an overcrowded class of frustrated students was not well-suited to Lizette's usual groupwork facilitation practices. Lizette described this realization later, recalling her thoughts at the time: "I can address scaffolding the questions differently. I can address the way that I'm going to pose questions, but I can't – I had not thought about how to completely change my strategies for teaching this class yet" (*Exit Interview, May 2019*). Thus, Lizette's understanding of her teaching as *helpful and supportive* was interrupted as she saw evidence that her pedagogical actions became a source of frustration, at least for the students in the video clip.

Describing what she saw of her teaching during the debrief, Lizette said, "Instead of doing something to help it click, I just kept walking around to the different groups and – I don't know [*sounding defeated*]." Ten minutes after Lizette made this comment, she left the debrief in tears, further confirming the painful shock of this realization. This reaction aligned with Elizabeth Self and Barbara Stengel's (2020) description of the relationship between negative emotions and "being pulled up short":

> We are caught off-guard, left reeling. We experience insecurity, emotional, physical, intellectual. The insecurity comes from what feels like an attack on one's self: cherished beliefs and traits we claim as our own are called into question. As we come face to face with our own limitations and blind spots, we are humbled. We experience loss.
>
> *(p. 65)*

Reaching the Limits of Lizette's Inquiry Stance

Lizette was a "problem-solving person" who embraced inquiry into practice. Yet during Fermat VFF 4, Lizette reached the limits of this stance for several reasons. First, she knew that the lesson did not go as she wanted, but the added scrutiny of the VFF increased the pressure she felt. As she told us later, "Things bomb, but never with people going back and listening to the tape over and over again from two different viewpoints of the whole ship burning" (*Exit Interview, May 2019*). She came into the debrief braced for a hard conversation, thinking she had already identified the underlying issue with the lesson: The content was harder than the students were prepared for, and the task was not clear.

Second, Katherine – who facilitated the VFF with Brette – was new to the research team and had only met Lizette once before. Throughout the project, we worked hard to establish trust and rapport with our participating teachers; in this case, Brette had worked closely with Lizette and Julie for all of Year 1. We had not fully anticipated the effect of bringing someone new into the VFF cycle, and we did not adequately protect Lizette in this vulnerable space. In her Exit Interview, Lizette disclosed how this heightened her emotions:

> That was the first time that I had Katherine [in a VFF debrief]. So it was a new person in the mix, too […] which puts a different pressure on it, right? Because I feel like you guys have talked a lot about how much you did like trust building, like those relationships.
>
> *(Exit Interview, May 2019)*

Being disappointed in her lesson was hard enough, but sharing it with a new person – with whom she had not yet established trust – amplified that difficulty.

Finally, Lizette described having "a really positive relationship" with the students in the video clip, so watching a clip where they were frustrated felt personal, like she let them down. Lizette was not resisting evaluation from us; she was resisting her own evaluation of herself as a "bad teacher." At this moment, Lizette was weighed down by the baggage of the binary – realizing the undesired consequences of her approach to groupwork made her feel like a bad teacher – leading to an emotional response. Once she had time to process her negative emotions, Lizette was able to move back into a productive space of co-inquiry. By working through this VFF, Lizette learned that:

> Even when it was hard and emotional, you guys have always really honored who we are and the hard work that we do, but also – whether it went well or not – pushed us to think about how to improve. So it's just been really helpful and changed my teaching for the better.
>
> *(Exit Interview, May 2019)*

What Supported Lizette's Learning to Learn

Normalizing Disappointing Lessons

In reflecting on Fermat VFF 4, Lizette cited this emotional response several times, describing vulnerability as a "really good thing, and also a really hard thing" for inquiry. She further elaborated what she learned about the experience:

> Vulnerability is big… You're going to get a lot of feedback, and it's going to be a mixed bag of feedback because some days are going to go well, and some days are not. And just being open to that is hard if nobody is ever giving you feedback. So, when the only thing you ever get is a pat on the back and a smiley face sticker, then when something doesn't go well and somebody is trying to help you with that, it's hard to receive in the beginning. I guess that's not my [whole] teaching practice. It's part of it.
>
> *(Exit Interview, May 2019)*

Although Lizette had described a desire for more authentic feedback when we started working with her, she learned throughout the study that authentic feedback can be difficult, especially when it is a new experience. Recognizing that difficult classroom moments did not indict her whole teaching practice, instead representing just "part of it," signaled Lizette's shift toward an expansive conception of good teaching, supporting her transition into a fuller inquiry stance.

Importantly, Julie also described realizing how vulnerable learning can be from witnessing Lizette's experience in Fermat VFF 4:

> [Watching Lizette's emotional response] prepared me for, "Okay, this is probably going to happen to me, but this happened to Lizette, too, so I'm fine." I didn't cry but I probably ... I felt a little bit afterwards like, "Okay, I need to take that all in." So I was ready, if they bring something up that I am not aware of, that's okay and that's normal and I think I wouldn't have been as prepared for that, because I don't think any of us had had quite a moment like that.
>
> *(Exit Interview, May 2019)*

Just as Abigail and Brad modeled enthusiastic engagement with feedback for Greg, Lizette modeled vulnerability for Julie.

Providing Additional Time to Respond to a Newly Diagnosed Problem

Although the "crying VFF" (as Lizette and our team came to affectionately call it) was a hard moment in our project, it also gave us a chance to demonstrate our commitment to Lizette's learning. After Lizette revised her diagnosis of the trouble with the focal lesson, expanding from *students struggled with the content and task* to include *students needed more resources for groupwork*, she worked with Katherine to brainstorm adjustments that would address this newly identified issue. They planned for Lizette to incorporate anchor charts – posters she could co-construct with her students – to provide additional resources as groups collaborated. This new practice of using anchor charts, which may initially seem superficial, helped Lizette reconceptualize what it means to *support students during groupwork*. Instead of support only coming from her direct assistance, she saw ways to build learning resources into the classroom environment. While the Fermat VFF 4 debrief was upsetting, it also spurred Lizette to seek new pedagogical actions to better align with her core values – the pedagogical responsibility tied to her teacher identity. Reflecting on the experience, Lizette said:

> One of the things I think that came out of that was, I've just done a lot more with how I work with my groups and how I set up those dynamics and less about how I set things up for the whole class – I thought about that a lot this year.
>
> *(Exit Interview, May 2019)*

By working through her disappointment and letting go of the baggage of the binary, Lizette was able to incorporate new practices to better support students' independence.

Veronica Kennedy: Moving Toward Expansive Collegiality

Throughout Project SIGMa, Veronica's participation in VFFs shifted from a stance of treating her colleagues as being evaluated to a stance of treating her colleagues as fellow explorers of uncertain and flexible practices. With this shift, Veronica brought unique perspectives and values to her collaboration with colleagues, prompting richer conversation and collective concept development across the school team.

Background and Context: Transitioning from Rees to Banneker

In Year 1, Veronica taught at Rees Middle School, where she had been teaching mathematics for about a decade. She had spent most of her career at Rees with her PDO colleague, Ezio Martín. Having been at Rees for so long, Veronica felt very comfortable with the core features of Rees's approach to mathematics instruction: shared problem sets, mastery-based assessments, teaching through productive struggle, and an emphasis on helping students translate between mathematical and English language. During the summer before Year 1, Veronica and Ezio attended a professional development program focused on groupwork, and they were eager to add groupworthy tasks to their repertoire – tasks that engage multiple mathematical abilities to foster student collaboration and interdependence (see Chapter 2). Ezio described Veronica as being "way ahead" of him in implementing groupwork, which Veronica attributed to her willingness to "take way bigger risks right away" (*Interview 1, September 2017*). This enthusiasm for experimentation was true to Veronica's personality; she quipped that she was always "the instigator and the push" among her friends, who frequently said, "We've got to reel her back in; she's too much right now." (*Interview 1, September 2017*).

Between Years 1 and 2 of our project, Veronica moved to Banneker High School, where she taught Algebra 1 and Geometry. She and her new colleagues – Abigail Graham, Clark Zapatero, and Franck To – used a scenario-based curriculum to teach Algebra 1, which had been written by Clark and Franck several years prior. The Banneker team committed to three shared pedagogical practices: (1) visual tasks to launch lessons; (2) visibly random group assignments; and (3) using vertical whiteboards to foster collaboration.

Where Veronica's Learning Started: As if her Colleagues Were Being Evaluated

Initially, Veronica responded to the VFFs as if her partner teacher, Ezio, was being evaluated. Like Greg's responses in the early Noether VFFs, much of Veronica's participation was focused on supporting Ezio, with her suggestions framed indirectly as wonderings. Generally, her co-inquiry moves expressed her alignment with Ezio as a colleague, invoking their collective perspectives and practices, and were decidedly non-evaluative. For example, in Veronica's first VFF (Rees VFF 1), Ezio had taught an eighth-grade lesson where students worked on the Tower of Hanoi problem (see Chapter 7).

During the debrief, after viewing a clip of students negotiating how to start the warm-up problem, Ezio observed that he had only heard two of the three students talking. Veronica asked, "Ezio, my question is, what would productive talk have looked like in that group of three? What would you have imagined they would have been talking about?" Rather than answering her directly, Ezio shared two concerns: First, one student seemed to be excluded from the group, and second, students disagreed about how to begin. He mentioned that the groupwork professional development had presented "very structured" group roles, whereas he had been "very loose" in his groupwork organization.

Veronica built on Ezio's observation, asking, "I'm wondering if there were more structured roles, would we have heard more conversation?" She explained what she had heard students saying in the video clip, diagnosing it as a situation where one student was directing the activity. She offered a strategy for explicitly encouraging collaboration, but then added "which I don't do either, Ezio, so I'm not judging you." Couching her suggestion in an explicit statement of alignment and non-judgment, Veronica communicated her resistance to evaluative talk. As the conversation shifted to other parts of the lesson, Veronica brought up group structures again: "I'm just wondering how you and I, when we go to the whiteboards, if we have some kind of procedure…" By consistently discussing Ezio's practice with the collective "we," Veronica made her questions about *their* practice, again mitigating evaluation in her statements. In Rees VFF 3, where Ezio was again the focal teacher, Veronica's participation continued along similar lines.

By contrast, when Veronica was the focal teacher in Rees VFF 2, she did not express any "wonderings;" instead, she simply noted what she did and added what she should have done or should do in the future. For instance, as the team watched a clip of her interacting with students, she started shaking her head as she watched. After the clip ended, Lani asked about Veronica's reaction:

LANI: What are you shaking your head about?
VERONICA: Because I feel like I didn't push them […] they had the pattern but then I did all of the talking.

Veronica continued elaborating on her disappointment that she did not give students an opportunity to share their thinking. As she summarized, "I shake my head because I used all the words. I didn't really push them to say it."

Veronica's markedly different form of participation in viewing her own classroom video seemed to lean into the binary, with its zero-point epistemology of best practices. Across the Rees VFFs, Veronica's engagement suggested that she interpreted the debriefs as evaluative, because the video clips illuminated instances where the focal teacher – whether Ezio or herself – had done something good or should have done something better. As a result, she responded to the debriefs in ways that supported and affirmed Ezio, positioning him as a good teacher. This careful conversational maneuvering showed sensitivity to her colleague and seemed shaped by the baggage of the binary. She was quick to directly compliment Ezio and to contextualize Rees so that we would understand why Ezio's pedagogical choices made sense; she couched her suggestions as questions to Ezio, swiftly aligning herself with him, suggesting that she would not have done anything differently. Yet when watching her own teaching, Veronica was critical of herself, rarely offering affirmations or explanations that supported her choices.

Moving to Banneker: Shifting VFF Protocols, Shifting VFF Stances

In Year 2, Veronica's move to Banneker High School introduced her to a new version of the VFFs. In the spirit of partnership work and co-design, Grace Chen and the Banneker High School teachers developed a modified VFF protocol to honor the team's preferences and needs. In Banneker VFFs, all the team members were filmed during each VFF cycle, so that, in a sense, they were all focal teachers; in this way, Veronica was a focal teacher in Banneker VFFs 4, 5, and 6. As was the case when she was the focal teacher at Rees, Veronica was critical of her own practice and did not pose many questions, instead asserting confidently what she could or should have done instead. With regards to her colleagues' practice, however, Veronica engaged in co-inquiry rather than taking the alignment stance she took with Ezio at Rees. As we elaborate, this shift was evidenced by her direct (instead of indirect) suggestions, sharing her own practice more than affirming others' practices, and questioning Banneker's taken-for-granted pedagogical practices.

More Direct Suggestions

At Banneker, Veronica made direct suggestions rather than framing them as wonderings. For example, in Banneker VFF 5, we watched a video clip from Clark's classroom in which he spontaneously asked a student to explain another student's whiteboard work. As the teachers discussed ways to make students more accountable for knowing multiple ways to solve a problem, Veronica shared a routine she had learned from Lee Bellver (see Chapter 7), saying, "It's another option for us to add to our whiteboard repertoires."

Later, we watched a video clip from Franck's classroom, where Franck attempted to invite a student named Roberto into a small group conversation by writing on the board near Roberto so that the other students in the group would have to turn and look at him. Franck presented this clip as a dilemma; he often found it difficult to integrate Roberto into groupwork because of the large gap between Roberto's mathematical fluency and that of his groupmates. Franck asked his colleagues, "What would I do, guys – team? Team!" After a few rounds of suggestions, Clark proposed, "What if we make him write?" and Veronica jumped in: "Yeah, make him hold the pen while [the other students] explain." Franck hesitated, saying, "But maybe that's like you forcing him to do something he doesn't want to do." Veronica acknowledged Franck's hesitation and continued to elaborate the hypothetical: "Okay, just as a brainstorm though…" She offered a possibility for how Franck might structure students' groupwork, adding, "It's tricky… yeah, no. There's a lot of holes in this brainstorm, but maybe…" By endorsing and then questioning different possible responses, Veronica reflected a shift toward a collaborative co-inquiry stance.

All four teachers continued to brainstorm ways to structure groupwork so that Roberto – and other students – could engage mathematically. After a few more minutes, Veronica introduced an additional factor to consider:

> There's more to school though… you have to think about the social context, too. It looks like he's participating, his body language is saying, "I'm doing what I'm supposed to be doing." And so maybe we're meeting a different need, which is allowing him to be near and participate in a low-risk way.

In contrast to her participation in Rees VFFs, Veronica took this conversation as an opportunity to probe possibilities – to test out, elaborate on, and perhaps reject different ideas – rather than as an opportunity to evaluate Franck's pedagogical choices or express support for Franck as a teacher.

Clarifying Shared Pedagogical Responsibilities: Articulating Dilemmas of Whiteboards

Throughout the Banneker VFFs, Veronica consistently raised questions about pedagogical practices that were shared across the team and the extent to which they aligned with their collective sense of pedagogical responsibility. We illustrate this through her continued questioning of a taken-for-granted practice among the Banneker team: having students solve problems in small groups on whiteboards hung around the room. For several years, Clark and Franck had students work on their scenario-based tasks in unstructured groups at vertical whiteboards. Arriving at Banneker in Year 2, Veronica and Abigail adopted similar practices. Although Veronica expressed interest in and commitment to whiteboarding – in fact, she had started exploring similar practices at Rees – she also experimented with variations on the unstructured groups that Clark and Franck seemed to prefer. For example, she once shared a variation where she paused her students and instructed them to take a "gallery walk" around the room to look at what other groups had written.

In Banneker VFF 4, each teacher taught an introduction to linear functions lesson in their Algebra 1 classes shortly after giving Unit 1 tests. A PDO leader had joined the debrief that day, and he shared his analysis of what was written on students' whiteboards at the end of each class period. He noted a striking consistency across whiteboards: Of the groups who had set up equations, they had all written the equations identically, even though many equivalent equations were possible. Furthermore, none of the whiteboards had any calculation errors. Building on these observations, the PDO leader asked about students "sharing" strategies across groups, which Veronica took as an opening to pose her own question:

> What I'm struggling with on the Unit 1 test is, like, the goal of the whiteboards is this mobility of knowledge. Everyone's going to be actively moving forward. And it feels as though, based on the Unit 1 test, not everyone gets the chance to move forward because perhaps they're looking for the answer. I don't know – that's where I'd like to brainstorm, because some of my students were clearly left behind in this unit.

Veronica worried that the ease of strategy-sharing across whiteboards might allow some students to appear to move forward by copying correct answers from others without actually learning the content. Because correct answers appeared on the whiteboards, however, she might have trouble identifying students who were still learning. In this way, she articulated a potential conflict between the team's pedagogical actions (an emphasis on whiteboarding) and their shared pedagogical responsibilities (supporting all students' mathematical learning).

Veronica continued to engage this tension – and to engage her Banneker colleagues with it – throughout the school year. In Banneker VFF 5, the teachers each taught a lesson in which students used an arithmetic series to determine the number of bricks in a wall. During the debrief, we viewed a video clip from Franck's classroom where a student hastily explained her whiteboard work to him. We discussed what the student likely meant in her explanation and how teachers might respond. Veronica remarked that the video clip reminded her of a similar conversation in one of her classes that had not been filmed:

> And I'm thinking, "Oh man, this context would be so much better in a number talk."[3] That kind of thinking is the kind of thinking you want

to share with everybody [...] Freeze frame and be like, "Okay, everyone come on over here for a number talk," [but] you can't. Our structure doesn't lend itself to that.

In this account, Veronica shared a perspective that the Banneker teachers' reliance on whiteboarding precluded the possibility of pulling the whole class together for a number talk, which could be useful in certain situations. By identifying a possible limitation of their taken-for-granted practices, Veronica gently prompted her colleagues to consider how they might want to modify their pedagogical actions.

Veronica continued to challenge the taken-for-granted nature of white-boarding at Banneker – both its unstructuredness and its dominance in most lessons – and, in Banneker VFF 6, she asked questions that pushed the Banneker team to explore and articulate what they expected to see while groups were working on whiteboards. Furthermore, she framed this discussion around the consequences of whiteboarding for different students in heterogenous classes. Through this discussion, Veronica created opportunities for her and her colleagues to clarify their values and commitments – their pedagogical responsibility. Veronica's participation in Banneker VFFs, especially when she was not discussing her own teaching, exhibited a strong exploratory stance and fostered the kind of dialogue that pushed her colleagues to articulate their pedagogical reasoning about their pedagogical actions in light of their peda-gogical responsibilities. In this way, she learned to embrace an expansive view of good teaching as a colleague in VFF conversations.

An Onto-Epistemic Shift: From Epistemic to Aleatory Uncertainty

Veronica's changing participation suggests an important onto-epistemic shift about the nature of knowledge in, of, and for teaching that goes beyond what we captured in Table 6.1. Namely, Veronica shifted from treating teaching as a problem of epistemic uncertainty to treating teaching as a problem of aleatory uncertainty. This is an important distinction: Epistemic uncertainty results from a lack of knowledge, while aleatory uncertainty results from the intrinsic variability of complex systems (Der Kiureghian & Ditlevsen, 2009). For example, teachers facing epistemic uncertainty might not know what to do in a situation because they lack sufficient information about individual students or particular conditions. They could, however, decrease epistemic uncertainty by learning more about these, thus increasing their confidence in their pedagogical choices. Aleatory uncertainty, in contrast, stems from the inherent complexities of the classroom. As a result, no amount of additional information would sufficiently account for all sources of variability to make a teacher truly confident that they were making good pedagogical choices.

With aleatory uncertainty, teaching becomes a wicked problem,[4] one for which, among other things, there is never a single solution and never a stopping point at which the absolute "goodness" of a pedagogical choice can be unimpeachably ascertained. Even with the same teacher, same students, and same class, what is "good" on one day might not be "good" on another day, or what is taken to be "good" after one analysis might reveal itself to have been not so good, in hindsight, several years later.

In Rees VFFs, Veronica's exploratory talk was cast with epistemic uncertainty, as her participation centered on assessing whether particular pedagogical choices were "good" in particular situations. Taking pedagogical choices to be "good" or "not good" invited evaluations of the focal teacher, which could make a caring and supportive colleague – like Veronica – feel the need to defend them. In Banneker VFFs, Veronica was a newcomer who shared the team's values but had a different teaching history. Her questions pushed further – approaching aleatory uncertainty – pointing to tensions and unknowability in the issues she raised. Veronica's participation enabled Banneker VFFs to become a space for open sharing, collaborative brainstorming, and exploration of pedagogical practices, values, and commitments, all of which supported the team's concept development.

What Supported Veronica's Learning to Learn

Colleague Characteristics

These shifts in Veronica's participation – making direct suggestions, sharing her own practice, and interrogating taken-for-granted practices – could be attributed to any number of conditions. Because the influence of these conditions is impossible to tease apart, we cannot claim that participating in VFFs produced these shifts. To name just some of the complexity: Abigail, Clark, and Franck were all different people than Ezio and had different personalities, histories, preferences, interaction styles, and interests. Veronica had different histories with them as colleagues and friends, and it is natural that Veronica would have worked differently with Ezio than she did with Abigail, Clark, and Franck. And Veronica's newcomer status at Banneker likely gave her an outsider-insider view that helped her make Banneker's familiar strange, possibly making some questions more salient to her.

Time and Comfort in VFF Settings

We also cannot disentangle the variables of time and place in understanding Veronica's changing participation. She participated in Rees VFFs during Year 1 of Project SIGMa and Banneker VFFs during Year 2; she may have become more familiar with the goals and possibilities of the VFFs and adjusted her engagement accordingly. Relatedly, Rees VFFs were facilitated by more senior members of the research team — Lani, the Principal Investigator, and Patty, the postdoctoral researcher — and Banneker VFFs were facilitated by graduate students. As a result, Rees VFFs may have felt more formal or high-stakes, and Banneker VFFs may have felt more familiar or casual.

School Support

Finally, as we alluded to earlier, Veronica saw herself as a skillful teacher at Rees, overseeing technological innovations in the mathematics department. However, Ezio and Veronica also felt at odds with their school leaders around instructional goals and priorities in teaching mathematics. In this context, where Ezio and Veronica both experienced heightened administrative scrutiny, Veronica may have felt it was too risky to critique their instructional practices. By contrast, the mathematics teachers at Banneker were strongly supported by their school leaders. This surer ground may have steadied any sense of threat and allowed Veronica to exercise her considerable analytic powers on her colleagues' teaching and on collective problems of practice, creating space to acknowledge, grapple with, and leave uncertainty unresolved instead of aiming to determine and implement what was definitively "good" or "best."

Discussion

In this chapter, we described how three teachers renegotiated subject-world relations. Specifically, they navigated the baggage of the binary, moving between conceptualizations of teaching as being clearly good or bad and conceptualizations of teaching as being highly situated, contextual, and uncertain. Although teachers rarely shifted fully and permanently from a binary onto-epistemic stance to an expansive onto-epistemic stance, teaching became discussable when they adopted the latter perspective. The expansive stance enabled teachers to integrate concepts about instruction, as their exploratory conversations pressed them to articulate the pedagogical reasoning for their pedagogical actions and hold this against their pedagogical responsibilities. By making teaching discussable, teachers learned to weigh the many different factors at play in any given situation and account for how those factors might influence pedagogical actions. This allowed them to revise their diagnoses of problems of practice, and the newly narrativized understandings supported reconceptualizations of practice. In other words, teachers were able to reason more ecologically when they inhabited an expansive onto-epistemic stance. As a result, VFF conversations where such a stance was evident contained more analysis, diagnostic talk, brainstorming, and learning, rather than evaluation and judgment. This, in turn, supported teachers' concept development.

Greg's pedagogical responsibility to be a good colleague may have supported his development of interpretive practice, helping him think differently about both the process of teacher learning and the details of student interaction in his classroom. Lizette's commitment to being a teacher learner gave her resources to draw on when she encountered a situation that pushed the limits of her inquiry stance, and normalizing disappointment allowed her to continue learning as she reconceptualized the problem with her lesson. Veronica's new context at Banneker offered her an opportunity to frame teaching as aleatorily uncertain, which created new opportunities for dialogue and experimentation, rather than simply identifying "best" practices. For all three teachers, the rich pedagogical reasoning made possible by taking an expansive onto-epistemic stance enabled them to align their pedagogical actions with their pedagogical responsibilities – their aspirations for their own teaching, for their students, and for their classrooms.

Greg, Lizette, and Veronica all felt like they were part of supportive mathematics teacher communities with great mutual respect between like-minded colleagues, both within their school mathematics departments and at the PDO more generally. Participating in VFFs and shifting to an expansive onto-epistemic stance on teaching made co-inquiry an integral way of being within their collaborative communities (Jaworski, 2006). In Greg's case, he initially believed he could learn from his colleagues, but that they would not have much to learn from him; by the end, however, he saw himself and his colleagues learning from observing and analyzing *his* classroom practice. Lizette modeled vulnerability for Julie, seeding her colleague's future co-inquiry and growth (see Chapter 8). And Veronica began to challenge pedagogical commitments that had become taken-for-granted in the Banneker community, enacting her conceptual agency as a teacher within the local practices. In these ways, teachers' renegotiation of subject-world relations through the VFFs served as an integral component of their learning to inquire into teaching.

Notes

1 Notably, Abigail never mentioned feeling offended during that VFF debrief. She reported that the discussion got her "thinking more about content [...] that I was kind of required to teach, and that I was wondering about the purpose of" (*Exit Interview, May 2019*). Of course, she may have shared a different sentiment with Greg and not with us. We may never know.

2 One early reader of this manuscript was skeptical of this claim, so we offer the following thought experiment: Imagine if instead of inviting us to their surest lessons, teachers invited us to their most challenging classes. The learning opportunities would be potentially greater, as would the potential payoff – getting insight into and identifying possible solutions for the challenges, making the class easier to teach. But the additional threat from the binary (*What if they think I'm a bad teacher?*) might not be worth the social risk. And given the uncertainty inherent in ambitious and equitable teaching, there was always plenty to talk about in VFF debriefs – even for relatively safe lessons.

3 A number talk is an instructional routine used by mathematics teachers to elicit and explore multiple ways of thinking about a problem (Parrish, 2010).

4 We thank Laura Carter-Stone for introducing this idea to us in relation to teaching.

References

Biesta, G. (2007). Why "what works" won't work: Evidence-based practice and the democratic deficit in educational research. *Educational Theory, 57*(1), 1–22.

Biesta, G. J. (2015). *Good education in an age of measurement: Ethics, politics, democracy.* Routledge.

Borko, H., Liston, D., & Whitcomb, J. A. (2006). A conversation of many voices: Critiques and visions of teacher education. *Journal of Teacher Education, 57*(3), 199–204.

Borko, H., Jacobs, J., Koellner, K., & Swackhamer, L. E. (2015). *Mathematics professional development: Improving teaching using the problem-solving cycle and leadership preparation models.* Teachers College Press.

Casey, L. (2012, March 1). The true story of Pascale Mauclair. Retrieved from http://newpol.org/content/true-story-pascale-mauclair

Chen, G. A., Horn, I. S., & Nolen, S. B. (2018). Engaging teacher identities in teacher education: Shifting notions of the "good teacher" to broaden teachers' learning. In *Research on teacher identity* (pp. 85–95). Springer, Cham.

Chen, G. A., Marshall, S. A., & Horn, I. S. (2021). "How do I choose?": Mathematics teachers' sensemaking about pedagogical responsibility. *Pedagogy, Culture & Society, 29*(3), 379–396.

Cohen, D. K. (2011). *Teaching and its predicaments.* Harvard University Press.

Coles, A. (2013). Using video for professional development: The role of the discussion facilitator. *Journal of Mathematics Teacher Education, 16*(3), 165–184.

Der Kiureghian, A., & Ditlevsen, O. (2009). Aleatory or epistemic? Does it matter? *Structural Safety, 31*(2), 105–112.

Ehrenfeld, N. (2021). Understanding Mathematics Teachers' Collaborative Learning in the Context of Teachers' Learning Ecologies. *Proceedings of the forty-third Annual Meeting of the North American Chapter of the International Group for the Psychology of Mathematics Education (PME-NA 43).*

Fenstermacher, G. D., & Richardson, V. (2005). On making determinations of quality in teaching. *Teachers College Record, 107*(1), 186–213.

Haraway, D. (1988). Situated knowledges: The science question in feminism and the privilege of partial perspective. *Feminist Studies, 14*(3), 575–599.

Horn, I. S. (2018). Accountability as a design for teacher learning: Sensemaking about mathematics and equity in the NCLB era. *Urban Education, 53*(3), 382–408.

Horn, I. S., & Kane, B. D. (2015). Opportunities for professional learning in mathematics teacher workgroup conversations: Relationships to instructional expertise. *Journal of the Learning Sciences, 24*(3), 373–418.

Horn, I. S., Garner, B., Kane, B. D., & Brasel, J. (2017). A taxonomy of instructional learning opportunities in teachers' workgroup conversations. *Journal of Teacher Education, 68*(1), 41–54.

Horn, I. S., Nolen, S. B., Ward, C., & Campbell, S. S. (2008). Developing practices in multiple worlds: The role of identity in learning to teach. *Teacher Education Quarterly, 35*(3), 61–72.

Jaworski, B. (2006). Theory and practice in mathematics teaching development: Critical inquiry as a mode of learning in teaching. *Journal of Mathematics Teacher Education, 9*, 187–211.

Kumashiro, K. (2012). Reflections on "bad teachers." *Berkeley Review of Education, 3*(1), 5–16.

Lemov, D. (2014). *Teach like a champion 2.0: 62 techniques that put students on the path to college.* John Wiley & Sons.

Little, J. (1990). The persistence of privacy: Autonomy and initiative in teachers. *Teachers College Record, 91*(4), 509–536.

Lovett, I. (2010, November 9). Teacher's Death Exposes Tensions in Los Angeles. *The New York Times.* Retrieved from https://www.nytimes.com/2010/11/10/education/10teacher.html

Luna, M. J., & Sherin, M. G. (2017). Using a video club design to promote teacher attention to students' ideas in science. *Teaching and Teacher Education, 66*, 282–294.

Parrish, S. (2010). *Number talks: Whole number computation.* Math Solutions.

Self, E. A., & Stengel, B. S. (2020). *Toward anti-oppressive teaching: Designing and using simulated encounters.* Boston, MA: Harvard Education Press.

Steeg, S. M. (2016). A case study of teacher reflection: Examining teacher participation in a video-based professional learning community. *Journal of Language and Literacy Education, 12*(1), 122–141.

Van Es, E. A. (2012). Examining the development of a teacher learning community: The case of a video club. *Teaching and Teacher Education, 28*(2), 182–192.

van Es, E. A., & Sherin, M. G. (2010). The influence of video clubs on teachers' thinking and practice. *Journal of Mathematics Teacher Education, 13*(2), 155–176.

Vedder-Weiss, D., Segal, A., & Lefstein, A. (2019). Teacher face-work in discussions of video-recorded classroom practice: Constraining or catalyzing opportunities to learn? *Journal of Teacher Education, 70*(5), 538–551.

Zhang, M., Lundeberg, M., Koehler, M. J., & Eberhardt, J. (2011). Understanding affordances and challenges of three types of video for teacher professional development. *Teaching and Teacher Education, 27*(2), 454–462.

7

LEARNING ABOUT TEACHING THROUGH MOMENTS OF INSIGHT

With Patricia Buenrostro and Samantha Marshall

As they gain experience, teachers learn to interpret and respond to various aspects of instruction. They learn to plan lessons, use different discourse strategies, and assess students' knowledge – topics frequently addressed in teacher education and professional development – but they also learn how to manage the flow of classroom activity, respond to unexpected questions, deal with conflicts, and so much more that goes beyond the scope of typical professional education. Given the breadth of things teachers learn to do, it is unsurprising that this learning – and what spurs and supports it – takes many different shapes.

Adding complexity to the phenomenon, when we study teacher learning, we note that the same outcome – say, implementing a specific discourse structure – proceeds along different learning trajectories for different teachers. For teachers who already use similar structures, their learning may be a matter of adding the approach into their existing repertoire; they do not need to do substantive conceptual work to understand the purpose of discourse structures to meet their pedagogical responsibilities. For those who perceive such structures as artificial – and thus conflicting with the teacher identities at the core of their pedagogical responsibilities – their learning may require them to experience a need for the strategy before reconciling it with their sense of authenticity.

These contrasting hypothetical learning trajectories highlight another source of variation in teacher learning: Different kinds of learning, for different teachers, take place over different timescales, even for teachers ostensibly learning the same thing. Sometimes, these timescales are brief. The right strategy offered at the right time can seed a new practice; a moment of

DOI: 10.4324/9781003182214-9

insight can yield new understanding. At other times, these timescales are longer, as teachers gradually refine a practice or engage in deep identity work as they reconcile themselves to something that initially seemed outside their comfort zone.

Video-formative feedback (VFF) cycles facilitated both kinds of learning, and most of our participating teachers experienced both brief moments of insight and longer learning trajectories as they refined concepts and practices over time. In this chapter, we highlight briefer moments of learning – critical events that catalyzed transformations in teachers' understandings or practices. Teachers' affect in these moments varied from the satisfaction of making new realizations (such as teachers listening to students make new connections) to the startling experience of being pulled up short (see Lizette, Chapter 6). We illustrate this insight-driven learning through the cases of Doha Arzoomanian, Ezio Martín, and Lee Bellver.

The Centrality of Facilitation as a Topic for Teacher Learning

As should be clear by now, ambitious and equitable mathematics instruction is challenging to enact. Centering student sensemaking through rich tasks and classroom discourse makes this form of instruction more uncertain than traditional mathematics teaching, as teachers are only partially responsible for how lessons unfold. Students often have unique insights or interpretations that even experienced teachers cannot anticipate. Teachers seeking to develop ambitious and equitable instruction can, of course, learn and refine techniques that support student sensemaking. Indeed, many of our participating teachers picked up new strategies from colleagues at the PDO, presenters at professional conferences, and even through their engagement on social media.

But learning about a practice – like asking students, "What do you notice? What do you wonder?" to orient them to a rich task – is not the same as knowing how to implement the practice in your own classroom, with its complex constellation of students. For instance, implementing the "notice-and-wonder" routine with students might raise questions like: *How do I explain to students what I mean by notice and wonder? How should I respond if students say something off-topic or non-mathematical? What if no one "notices" what I want them to see?* Depending on the task or context that students are asked about and the mathematical content that teachers hope to elicit, more specific questions might arise.

To reiterate a core point, responsive instruction requires that teachers develop integrated concepts about instruction. Knowing *about* a strategy is insufficient; teachers must also learn about *why, when,* and *under what conditions* a strategy is useful, as well as how to respond to the inherent uncertainties of

TABLE 7.1 Overview of Doha's, Lee's, and Ezio's learning.

	Doha	Ezio	Lee
Where they started	Expressed desire to involve more students and with more conceptual questions in whole-class discussions	Expressed desire to support students to develop perseverance in the face of mathematical difficulties	Expressed desire to facilitate student sensemaking so that students could see themselves as authors of mathematics
Critical moment	After Doha explained her reasoning for interrupting small groups or not, Lani introduced the "huddle" strategy	After observing Ezio's repeated interactions with small groups, Patty suggested a scaffolding strategy to better support students' agency in mathematical sensemaking	After analyzing his interaction with a small group, Lee realized he had only provided them with procedural resources, not conceptual resources
What made this a critical moment	• Identified misalignments between pedagogical responsibility and pedagogical action • Opportunity to brainstorm alternative approaches	• Connected the suggested scaffold as the opportunity to facilitate students' articulation of the task structure • Identified chunking as a way to support multiple goals	• Line-by-line analysis of students' work and sensemaking • Additional perspectives and possibilities shared by colleagues

classroom interaction. As we described in Chapter 5, most of the problems of practice that arose in VFFs related to issues of facilitating classroom interaction, underscoring the complexity of this aspect of ambitious and equitable mathematics instruction. In this chapter, we narrate critical moments in Doha, Ezio, and Lee's learning that were supported by VFFs. A summary of their cases can be found in Table 7.1.

Doha Arzoomanian: Supporting Students' Mathematical Agency

Background and Context

At the start of our study, Doha Arzoomanian had taught mathematics for over two decades, including many years at Falconer Middle School. She began teaching in her home country and, after moving to the United States, taught at a school near Falconer. Doha's partner teacher, Lee Bellver, recruited her to work at Falconer. Lee told us that when he interviewed teacher candidates, he would ask, "Do you like children?" and pay close attention to interviewees' responses. Spending time in Doha's classroom and watching her interact with students, we can easily imagine how she would have smiled and told endearing stories about her students in response; her affection for their mischief and humor was unmistakable. After her hire at Falconer, Lee and Doha quickly became close friends and colleagues, often co-planning and observing one another. Their administrators trusted them and largely let them teach and experiment as they saw fit.

Our data from Falconer are slightly different from those at other schools. Samantha Marshall, the team's primary research contact, secured additional funding to work with the Falconer teachers more intensively, allowing her to spend twice as much time with them as we spent with other school teams. Sammie also modified the VFF process, co-planning lessons with Doha and Lee before each VFF cycle, seeking to support their learning about instructional design. Some of these extended cycles led to more discussion about mathematics content, conceptual understanding, task design, and planning instructional sequences, but in the debriefs, our PoPs analysis (see Chapter 5) showed that the Falconer team also had substantive discussions about interactive elements of teaching.

Doha's Inquiry: Whole-Class Discussions

Throughout the project, Doha raised co-inquiry questions about improving her whole-class discussions. Starting with Falconer VFF 2, Doha asked how she could improve her whole-class questioning. In Falconer VFF 4, she asked, "How can I keep and engage more students during whole-class discussion?" Putting a finer point on her concern, in Falconer VFF 8, she asked, "How can I avoid turning the whole-class discussion into a lecture?" The consistency of her questions about whole-class discussions signaled that this topic had cognitive salience for her. Not surprisingly, then, the Falconer team talked about whole-class discussions relatively often, sometimes in conjunction with other ideas, like supporting students' conceptual understanding, scaffolding and posing questions, and even the challenges of managing a large class.

Where Doha's Learning Started

Doha's relationships with her students were marked by gentle affection. She often smiled warmly at students as they came in and playfully redirected them when they were distracted. She relished regaling us with stories about her students, whom she got to know and love.

When we first began observing Doha, we noticed her tendency to hurry through whole-class discussions. Doha told us that her purpose for whole-class discussions was typically to clarify students' misunderstandings, but the conversations often felt rushed, likely because she was pressed for time to get through her lessons and because she had large classes, with up to 45 students. From our perspective, it seemed that only a few students were able to follow along and consolidate their learning during these discussions. Part of her challenge was facilitation: She spent a great deal of time (lovingly) shouting to get the class's attention ("Guys, guys, guys... I need your attention up here!") as she tried to gather a room crowded with middle-schoolers. In her whole-class discussions, she often asked known-information questions and used "equity sticks" – popsicle sticks with students' names – which she picked at random to distribute participation.

As Doha worked with Sammie and other members of the research team, we discussed various elements of leading whole-class discussions and shared strategies she could implement in her classroom. Specifically, Doha wanted to involve more students in these discussions. We considered strategies for broadening participation by giving students opportunities to build confidence in their thinking, such as think-pair-share, where students first think about a question independently, then discuss with a partner before sharing with the whole class. This was important for many of Doha's students who, like her, were multilingual and potentially self-conscious speaking in front of large groups. Doha also wanted to ask more conceptual and information-seeking (rather than known-information) questions.

Doha's Critical Moment in Falconer VFF 6: Using Huddles to Support Student Discourse

In Falconer VFF 6, Doha had an especially meaningful insight – what we characterize as a critical moment – that transformed her practice. Sammie and Lani facilitated the VFF debrief; Lee was out of town. As we will show, Lani suggested a particular strategy ("the huddle") that resolved an ongoing dilemma in Doha's facilitation of whole-class discussions. This quickly became one of her stock strategies for facilitating student discourse. In both the Exit and Member Check Interviews, Doha described the Falconer VFF 6 debrief as being an important moment in her learning.

Setting Up the Inquiry

Like many of our participants, Doha drew on the TRU Framework (Teaching for Robust Understanding; Schoenfeld et al., 2020) to develop inquiry questions. The TRU Framework is a research-based rubric that delineates five essential dimensions of ambitious and equitable mathematics instruction.

These are *content; cognitive demand; equitable access to content; agency, ownership, and identity;* and *formative assessment.* The teachers first encountered the TRU Framework at a PDO workshop in Year 1. Many of the PDO Working Groups – including Doha's – used the TRU Framework to guide their monthly conversations.

Doha planned to share video excerpts from her VFF lesson at her next Working Group meeting. Her co-inquiry questions, then, reflected the group's focus on the TRU Framework:

1. How can I avoid the whole-class discussion turning into lecture?
2. How can I improve with regard to the TRU framework, specifically, improving students' "perseverance" (and avoiding scaffolding too much)?

These questions pointed to Doha's sense of pedagogical responsibility – to let students make sense of mathematics as they persevere through complex problems and discuss their strategies as a class – which she wanted her pedagogical actions to reflect.

Lesson Overview

In Falconer VFF 6, Doha's 8th-grade Algebra 1 students were learning about properties of exponents. Her goal was for students to develop rules for simplifying exponent expressions and to understand the meaning of negative exponents. The lesson began with a warm-up, in which students used the expanded form of exponents (e.g., $x^4 = x \cdot x \cdot x \cdot x$) to explore the meaning of ratios like x^4/x^6 and identify patterns in multiplying and dividing exponents. After about 30 minutes of small-group work on these exercises, Doha called the class back together for a brief discussion. Then students moved on to practice problems.

As she circulated around the room, Doha noticed many students struggling with a problem that asked them to simplify $(2x2y)^3(-3xy^3)^2$. She interrupted the groupwork, calling the whole class together to discuss this problem. A few minutes later, she noticed another widely challenging problem – $(3xy-2)^2(2x3yz)(6yz^2)^{-1}$ – and again interrupted the groupwork to discuss it. For the remainder of the lesson, students worked in small groups as Doha circulated and answered questions.

Deciding When to Interrupt Small-Group Work

Related to Doha's co-inquiry questions, the issue of when to interrupt students' groupwork for whole-class discussions emerged during the debrief.

The lesson had three whole-class discussions: one following the warm-up and two interruptions in response to difficult problems. Student involvement was especially uneven in the whole-class discussions that came from interruptions.

When the debrief began, Doha told us that she was reluctant to spend too much time in whole-class discussions since she was trying to avoid lecturing and instead wanted students to take on more ownership and agency. She said, "I was thinking about this, about the access, about the ownership, the kids. When we discussed before, that was my worry. Sometimes I take over too much... Sometimes I scaffold too much."

Sammie noted that Doha spent much of the groupwork time asking groups the same question: "I saw you walking around asking every single group the connection between the negative exponent and the reciprocal." With this comment, Sammie wanted to press Doha's pedagogical judgment, asking her to consider when it might make more sense to talk to the whole class and when to check in with small groups, especially if she found herself repeating the same question. Doha agreed that it was difficult to check in with each group about the same issue. While this would be true in most classrooms, it was especially true in Doha's: There were 44 students that day, so Doha felt like she had to "run around" to get to them all. At the same time, Doha was reluctant to pull the class together too often, since she tended to launch into monologues (or "lectures") – as she described it, "I carry on. I don't stop."

When Sammie introduced video clips, she framed the focus of the conversation as watching the lesson and thinking about how Doha could re-imagine her instruction – a way of brainstorming alternative approaches through collective interpretation. In the first clip, Doha asked students to evaluate algebraic expressions; after watching this, she echoed her earlier sentiment: "I model too much." She described one student, Cindy, who had found an especially creative strategy for simplifying exponents. Doha asked us, "What do you guys think? What if I asked Cindy to do it on the board for everyone?" As we discussed the possibility of bringing students to the board, Sammie asked Doha why she thought that would help. Doha tied her pedagogical reasoning to a formative experience that she had in elementary school:

> It's confidence. It's intimidation. All that stuff. [...] You feel like, "You know what? My teacher checked it was correct. I'm going to go show this to my classmates. I will get maybe an extra point for this." When they go do it on the board, they retain that for longer. I remember until now, I wrote a problem on the board, in 4th grade. I was in 4th grade, I remember. We were just trying decimals. My teacher said, "What's the

number between seven and eight on a number line? Write the number."
I remember, I swear, I remember this. I went to the board. I put, it's
seven, and there should be like it's seven. When you have money, it's
$7.50 and then, he said, "Put five, you're right." [...] And we were just
learning decimals. I swear, I remember this. I went there, I wrote – you
know, we had blackboards. I wrote with chalk and I – I never forgot
this problem.

As Doha described this significant experience from her childhood, she suggested that it might similarly boost Cindy's confidence if she were invited to
share her ideas at the board. Sammie connected this to the TRU framework
and how it could support Cindy's – and other students' – ownership and identity in mathematics.

SAMMIE: Students have opportunities to conjecture, explain, make mathematical arguments, and build on one another's ideas. And I was thinking
about if you did have her up at the board, this would be a really cool
opportunity to explain her conjectures and make an argument, but it
would also be an opportunity for, then, other students to build on her
ideas.
DOHA: Absolutely.
SAMMIE: So if she put her work up but then you could either have her explain
it and this is all – There's not a right decision. You could have her explain
it, or you could say, "Someone else explain what Cindy did."
DOHA: That's – that's even better. I think. I don't know. If someone else can
explain what she did, more learning for more people, more discovery.

In this last turn, Doha provided a rationale for this alternative practice of
having one student show their work at the board while another explains that
student's work – "more learning for more people."
    Lani then asked why Doha did not have Cindy explain her thinking at the
board. Doha explained that she was worried about the time it would take,
and knowing her student, she was concerned that Cindy would be afraid to
talk, especially on camera. These considerations reflected Doha's pedagogical responsibilities to get through a lesson and respect students' comfort,
respectively.
    Lani suggested that if a student were shy, Doha might "have them explain
it to a partner and have them go up together to the board and explain it
together," elaborating:

I've had entire groups go to the board sometimes to present something,
even though mostly the thinking is coming from one kid. So, I'll say
to Cindy, "Make sure your whole group understands, and then you all

can go to the board together." Again, that's more time – but I would do that, maybe, if I really wanted Cindy to have a chance to share her idea and shine. There's a lot of ways you can scaffold the participation part of it, too.

During our Member Check visit in Year 3, we saw Doha use this exact practice, bringing several students to the board to explain their thinking. We asked her about it, and she confirmed that she did this on nearly a daily basis, starting at the beginning of the year. She also shared that sometimes she checked students' work before they went up, since some students were afraid to make mistakes. This showed her characteristic concern for students' feelings as she supported their ownership of mathematical ideas.

In the VFF debrief, we went on to discuss a second clip, considering issues of when and how to press on students' answers in support of their thinking. We discussed the importance of thinking ahead about the different problems in a practice set, noting that some were harder but nonetheless important for students' sensemaking. Lani used the metaphor of "steepness" on a walk: When the road gets steeper, it takes more exertion, but it also builds muscle. Similarly, when students encounter "steep" parts of the lesson, it may be more difficult for them to make progress, but it also can pay off in improved understanding. We then brainstormed how to support students through those steep parts. This discussion tied questions of where to interrupt to issues about the quality of the mathematics students were working on, as the work might look different depending on the steepness of the problem.

We then watched a third clip, which showed Doha bringing the class together for her first interruption after she noted widespread student confusion. She was giving a mini-lecture, during which she posed some conceptual questions (e.g., "What happens when you expand a^3? Is there a connection between this and a times a?"). Doha was very critical of herself in viewing these clips, feeling frustrated that she did not elicit more student thinking. We discussed the dilemma she felt between keeping students in small groups and repeating herself to each group – a challenging approach with such a large class – versus interrupting them to help them through the steep part of the lesson.

Learning About the Huddle

While deliberating this dilemma, Lani asked Doha to explain how she had organized small groups and what expectations were in place for groupwork. Doha explained strategies she used, like random grouping and giving students different colored markers. Sammie pointed out that when Doha circulated to groups, she usually helped one student at a time. Doha agreed, saying that normally, she would circle something on a student's paper, and "then the conversation happens between two of us." Sammie asked about the possibility of setting expectations that students talk with one another in their small groups; Doha agreed this would be a good change. Doha then imagined giving each group small whiteboards to show their work, allowing for more group discussion.

Lani reflected with Doha about how that might solve her dilemma:

LANI: So [*the whiteboard strategy*] it's more table by table? Instead of having to bring the whole class together? I just think those judgments, where you saw something, and you went around the tables – this is what Sammie was saying earlier. That's a formative assessment moment. Then you realize, "I'm losing some kids here," so you bring the whole class together. That's one option, is to bring the whole class together. But I think part of what Sammie's trying to think about is, what are other strategies for getting people together? [...] One is to bring the whole class together and do a mini-lecture kind of thing like you did. [...] You decided another one, which is, have whiteboards and do it group by group, and you see, [one student] gets it, [a second student] is struggling. "[First student], could you explain it to [second student]"?

DOHA: That's good.

LANI: Right? So it's more group by group, so you don't have to stop everybody's work. That's the trade-off.

DOHA: I think that's going to work in that class that way.

LANI: Especially with 44 kids. If they start to learn to talk to each other more and use each other as resources more, then they're not turning to you every time.

Doha agreed that this strategy would be helpful; she wondered aloud about how it would work in her classroom. She described pulling half the class up to the boards while the other half worked at tables. Lani suggested another alternative:

LANI: When you notice that happening, where they're in different parts [of the problem], instead of interrupting the whole class, you can say, "Send one person from your group up" and do a little huddle, so that the rest of the group can keep working, but you can have a key question. The question you asked about, right? You could, instead of doing that for the whole class, you could sort of send a delegate from your group [...].

DOHA: Oh yeah, I have that too, yeah. I have ... Okay.

LANI: So you could just have one person come up and have a little huddle so at some part of the class ... Everybody else is still working. The delegate from the group, "So, you guys, this is a really important idea: a^3, how are you thinking of what a to the third power means?" And have that little conversation. "Now, go back to your groups and make sure everybody understands." Again, thinking about this idea, because then the kids are responsible for explaining it to their peers.

DOHA: That's so important.

LANI: Well, I'm thinking that this is one of the things you're wanting to think about.

DOHA: You know, that's going to work.

LANI: I think your relationships are so amazing –

DOHA: This is going to work.

LANI: Yes, that I think, I agree, that it will work.

DOHA: This is a really good idea. One person from each group, so what, 10 kids? 10, 11? Then they go back and explain it to their group.

LANI: So it just depends –

DOHA: The rest are working, the other three kids who stayed, they just keep working, whatever they were doing?

LANI: Yeah, yeah.

DOHA: And then one kid, he knows for sure, he goes back.

LANI: These different kinds of interruptions. This is, again, a common problem. The groups are in different places and you see –

DOHA: You know what? I'm going to try this, guys, and then next time, when we video, let's see if it works.

Doha clearly saw the huddle – which balanced the tension between giving students just-in-time support with her desire to neither lecture at them nor interrupt their groupwork – as a viable strategy. Her face lit up as she enthusiastically repeated that it was "going to work." In fact, she saw it as solving a regular challenge in her classroom:

DOHA: That's going to work perfectly in my other class. Sorry, guys, I'm just thinking about what classes I have. Managing is going to be easier. I'm going to practice this and then I'll tell you later.

LANI: Sometimes interruptions are good, and you need to do them.

DOHA: Yeah, but not in this case.

As Sammie pulled up another clip, Doha exclaimed, "With different eyes I am watching this, guys, with different eyes with you. It helps."

Using the Huddle Later

The huddle became a favorite strategy of Doha's – in her Member Check interview, she mentioned "with larger classes, I still do the huddle a lot." Later in the interview, she described how much better her whole-class discussions were going, saying that she had a better sense of when to stop. She also riffed on the practice, asking students who had finished to lead other groups – what she called "mini-huddles" – to manage her large classes: "I said, 'Guys, this is your group; this is mine.' I took the large one, and they took three and three. I do that a lot. Huddle, mini-huddle kind of thing."

In Falconer VFF 7, Doha framed her co-inquiry question around the huddle – asking how it went, and how she could improve it. In that VFF cycle, we found that the huddle itself went well, but that as students returned to their groups after the huddle, they just gave their group answers instead of explaining their thinking. In that debrief, Sammie helped Doha strategize ways to support students to share with their groups after a huddle, thus helping her refine the practice. Although Falconer VFF 6 was a critical moment, it was also the start of a longer trajectory of her learning.

What Made This a Critical Moment: Seeing with "Different Eyes"

The Falconer VFF 6 debrief addressed a persistent problem in Doha's practice – how to manage moments when she noticed students getting off-track with the mathematics, while managing both time and her desire to foster their mathematical understanding. Her continued use of the strategies introduced in VFF 6 show that the conversation had high relevance for her learning. As her co-inquiry questions suggested, she wanted to foster students' agency and ownership of the mathematics they were learning, avoiding unnecessary lectures, and supporting their perseverance through challenging problems.

In our facilitation, we strategically and persistently pointed out misalignments between her pedagogical responsibility and pedagogical actions. Because Doha drew on the TRU framework and expressed a commitment to developing students' ownership and agency, we were able to focus on moments of missed opportunity captured on video, helping her re-imagine other ways of working in the future. For example, she could see that when she did most of the talking during whole-class discussions, students were not given opportunities to own the mathematics – a clear conflict between her pedagogical responsibility and pedagogical actions.

We also pressed Doha to consider the unintended consequences of how she was interacting with student groups: By helping students individually, she was not giving them many opportunities to help one another and work collaboratively. The huddle, along with the other strategies we discussed, helped her bring her pedagogical actions into greater alignment with her pedagogical responsibilities, contributing to new understandings of and practices for supporting student ownership and agency. This conversation helped Doha integrate the formal concepts of the TRU framework with lived concepts of her classroom. As a result, she saw her classroom "with different eyes," helping further her learning about student ownership and agency of mathematics.

Ezio Martín: Encouraging Productive Struggle

Background and Context

When our team met him, Ezio Martín had over 20 years of teaching experience in two different middle schools. During Year 1 of Project SIGMa, he taught at Rees Middle School with Veronica Kennedy.[1] Mathematics was a core part of Ezio's identity both within and outside the classroom. He held multiple degrees in mathematics, decorated his classroom with posters of mathematicians, and lined the chalkboard trays with books celebrating the joy of mathematics, including titles like *Alice in Puzzle Land* and *Magical Mathematics*. He wanted to impart mathematical culture to his students, asking them to learn the names of at least ten mathematicians by the time they left his class.

Ezio also identified strongly with working-class Latinx communities. At his previous school, after he noticed that upper-track courses had lower Latinx representation, he and a colleague collaborated to increase Latinx students' enrollment in advanced math classes. He left that school after almost two decades, disillusioned with declining enrollments, changing neighborhood demographics, and increased charter-school presence. During Project SIGMa, similar issues were emerging at Rees, with unofficial academic tracking and an explicit, school-wide focus on STEM as school administrators attempted to recruit more affluent families. In conversations with us, Ezio frequently revisited these issues of tracking and gentrification since they deeply troubled him.

For Ezio, a core pedagogical responsibility was to personally make a difference for his students. He wanted to know that his instruction and personalization mattered to students' academic success. His interactions with students were playful yet demanding. He was caring and compassionate and, according to Veronica, many former students came back to visit him after moving on to high school and college.

Ezio's Inquiry: Helping Students Persevere and Overcome Challenges

Ezio's co-inquiry questions reflected his pedagogical responsibility to help students develop perseverance in the face of challenging tasks. He connected this to one of the Common Core Mathematical Practice Standards: *Make sense of problems and persevere in solving them.* Based on his experiences both learning and teaching mathematics, Ezio had developed a personal theory that every student eventually hits a "wall" or "breaking point" when they encounter a difficult problem or course. Some students, Ezio explained, reach that point and feel like they can no longer continue in mathematics. Others, however, learn to persist through the difficulty and "overcome that breaking point," which allows them to recognize their strengths and continue studying mathematics. To help students develop perseverance – the skills to overcome those inevitable breaking points – Ezio provided them with complex problems to intentionally push them toward a breaking point. This, in turn, would give him an opportunity to support them in working through challenges and developing perseverance. Ezio felt that these experiences – facing a challenging task, struggling to make sense of it, and ultimately finding the solution – would help students see themselves as competent and capable mathematicians, setting them up for success throughout their academic careers.

Where Ezio's Learning Started

When we first visited Rees in Year 0, Ezio and Veronica described a goal of using low-floor, high-ceiling tasks – tasks that have multiple entry points and give students access to rich mathematics (see Chapter 2). Specifically, Ezio wanted to foster students' agency through their direct exploration with mathematics. In our theoretical language, Ezio wanted to make an onto-epistemic shift in his organization of classroom activities to better reflect ambitious and equitable mathematics instruction.

Ezio was also wondering about students' dynamics during groupwork. He had recently begun implementing cooperative learning, and he wanted to ensure equitable participation within groups. Both Ezio and Veronica were concerned about unofficial tracking practices that had emerged at Rees, with affluent and working-class students getting placed in separate sections of the same courses. They were especially worried about the related messages that students received about themselves as mathematics learners, tying this to the context of gentrification. Stating it plainly, Ezio raised concerns over the visible-yet-not-official tracking in mathematics and, relatedly, students in the "low" tracks thinking they were part of the "dumb" class.

Ezio's Critical Moment in Rees VFF 1: Trying to Do Too Much Too Fast

In Rees VFF 1, Ezio presented his students with The Tower of Hanoi, a classic mathematics puzzle. Before filming, Ezio expected that students would struggle with the task, reflecting his goal for students to persevere. Afterwards, though, Ezio was dissatisfied with the quality of their struggle, feeling that he could have better facilitated students' engagement in the task. He sensed that students had struggled too much with the first part of the activity – solving the puzzle – prompting him to offer students a hint, which he called "the cheat." The cheat helped students get to the second part of the activity – generalizing their findings – but without understanding where the cheat came from or why it worked.

During the debrief, Patty offered the heuristic of *solving a simpler problem first* as a scaffold for Ezio to support students' exploration of the puzzle's structure. Ezio made an important realization that the scaffold could have been instrumental in students' discovery of the "cheat" for themselves, supporting their sensemaking and meeting his goals of showing them that they could "overcome the breaking point." In this way, discussing strategies to support students' perseverance helped bring his pedagogical actions in line with his pedagogical responsibility. Ezio realized that he was trying to do "too much too fast," and that by "chunking" the lesson – or extending the task over several days – he could better support students' appreciation of the mathematical richness of the Tower of Hanoi.

Lesson Overview

Ezio introduced the Tower of Hanoi by presenting students with physical models of the puzzle and handouts with instructions. He explained how the puzzle worked: The goal is to move a stack of rings (see Figure 7.1) to a different pole by only moving one ring at a time and without putting a bigger ring on top of a smaller ring. He then instructed students to work in groups to figure out the minimum number of moves to solve the puzzle with various numbers of rings and to record their findings on the handout. Once students found the number of moves to solve the puzzle with up to seven rings, Ezio prompted them to identify a pattern and find a general rule for the minimum number of moves to move n rings.

As students worked, Ezio circulated among the groups and asked them to articulate how they found the minimum number of moves for a given number of rings. Some groups solved the puzzle with the minimum number of moves, while others found less efficient solutions. Since the *minimum* number

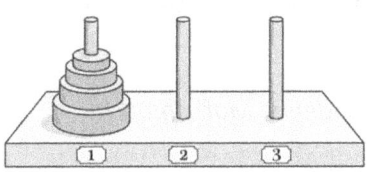 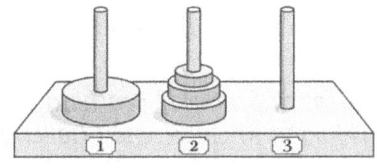

FIGURE 7.1 Illustration of the Tower of Hanoi puzzle with four rings.

Note: The puzzle's starting point is on the left, with four rings on Pole 1. On the right, the puzzle is at the "halfway point" (as Ezio referred to it). It has taken seven moves to get the top three rings to Pole 2. It will take one move to bring the bottom ring to Pole 3, and then seven more moves to bring the top three rings to Pole 3, which makes 15 moves to solve the four-ring puzzle.

of moves is necessary for generalizing the pattern for n rings, Ezio challenged each group with the same question: "How do you know that [your number] is the minimum?" Notably, Ezio did not follow up with groups after asking this; instead, he continued circulating around the room. Most groups correctly found the minimum number of moves for one, two, and three rings. But when they worked on solutions for four or five rings, groups struggled to find the minimum – or, at least, to feel confident that their solution was, indeed, the most efficient.

Ezio wanted to help students move past the counting phase so that they could look for the general rule. In the last 30 minutes of class, Ezio began showing groups how to use his "cheat," using the three-ring solution to find the four-ring solution, and then extrapolating from four rings to five.

The "Cheat"

Here, we take a brief detour to explain the cheat. Ezio's cheat drew on recursive logic: The solution for one ring (s_1) is one move. The solution for two rings (s_2) is three moves – move the top ring to Pole 2, then the bottom ring to Pole 3, then move the top ring onto the bottom ring. The solution for three rings (s_3) can be thought of as the solution for two rings (three moves to get the top two rings to Pole 2), one move (to move the bottom ring to Pole 3), and then the solution for two rings (three moves to get the top two rings onto the bottom ring) – or, $s_3 = 2s_2 + 1 = 7$. Similarly, solving the puzzle with four rings involves moving the top three rings (seven moves), then the bottom ring (one move), then the top three rings (seven moves), for a total of 15 moves. Using this pattern, it is possible to find the minimum number of moves for any number of rings (s_n), given the previous solution (s_{n-1}): $s_n = 2s_{n-1} + 1$.

This recursive pattern was Ezio's cheat. Importantly, the cheat does not reveal the explicit rule that Ezio wanted his students to find, since each solution relies on the previous one – that is, finding s_n requires s_{n-1} – but it would generate enough correct values to help students see a pattern and derive the formula to directly calculate s_n.

Cheat or Scaffold? Learning to Support Students' Mathematical Thinking to Overcome Challenges

Recall that Ezio's inquiry questions for the VFF were around student perseverance and the quality of their groupwork. During the debrief, Patty shared the general patterns we had observed as he circulated among groups. As we noted, Ezio asked the same question to each group – about their confidence that they had found the minimum number of moves – but did not follow up with them. To illustrate this, Patty shared a clip of Ezio approaching a group. After watching the clip, Ezio responded:

EZIO: Right, I threw that out, but I never asked them. No one had shown me that they knew it was [the minimum number of moves]. Oh, that was the one thing I was hoping. As they were moving their rings, I didn't hear anyone say, "Well, I've reached the halfway point, so it shouldn't be more than double that." I didn't hear anyone making that conclusion.

PATTY: Yeah, I think what we heard a lot when we listened to the different markers[2]…besides a lot of clicking and clacking of the rings, the rods. It was a lot of counting, right? "Let's try it again: one, two, three. How did you do that? Let me try." Students didn't have a way to talk about the things that you [are] raising, right?

Ezio had hoped that his question would help student recognize the pattern that consistently leads to the minimum number of moves – and that they would describe it as he saw it, with the "halfway point" alluding to the recursive doubling-plus-one pattern.

After repeated trials, students correctly reproduced seven and 15 moves as the minimum for three and four rings, respectively, but their talk ("Let's try it again," "How did you do that?") showed that they were using trial-and-error and not analyzing their solutions or looking for patterns. Veronica noted the challenge of getting students to "discuss and verbalize their thinking." She and Patty recognized the thinking that students were clearly doing in the video; as Veronica asserted, "they're coming up with strong conclusions […] Ezio gave a probing question […], but it's that next level of verbalizing the why."

Seeking more specificity in Ezio's lesson design, Lani asked him to describe the kind of talk he expected to hear:

LANI: What would it look like for this problem, the Tower of Hanoi problem, given the content that the kids know so far? What would be an example of them explaining their thinking and verbalizing their thinking in a way that you would hope to see?

EZIO: Again, I go back to that – I was hoping I would've heard someone say on that last ring, as soon as they're moving them, "Okay, that's my 15th move so it should be more than 30."

PATTY: The cheat. You wanted them to kind of start to articulate the cheat.

EZIO: Kind of, the early stages of the cheat.

Veronica reiterated the challenge with this goal: "The same thing, my kids were doing good thinking, but they weren't saying it out loud." In other words, Ezio's expectation did not reflect behaviors that students consistently demonstrated. And he was not alone in this – Veronica's students had similar difficulties verbalizing their thinking.

Shifting from a "Cheat" to a "Scaffold"

As Ezio explained his rationale for introducing the cheat – moving students forward so they could work on the generalization – Patty suggested an alternative way to meet the same goal. She proposed that, instead of giving students a "cheat" to find the solution more quickly, a *scaffold* might help students figure out the solution for themselves. The cheat and the scaffold rely on the same underlying reasoning – drawing on the recursive pattern and using a known solution to find the next solution. But by framing it as a scaffold, Ezio could support students to develop the strategy themselves, rather than directly giving it to them.

Patty explained that a problem-solving heuristic – *solving an easier problem first* – could be a useful scaffold for understanding the Tower of Hanoi. Building on students' solutions for simpler cases, Ezio could have encouraged them to analyze their moves and, as a class, collectively articulate the pattern and extend it to more complex cases.

PATTY: One thing I'm thinking of is, if you take the simple one – three rings produce seven moves, right? That's the minimum number of moves. Asking students to duplicate it, right? Really, they understand the pattern. They understand the pattern.

EZIO: Right.

PATTY: Once they're positive, "I can do it in seven [moves] as many times as you want me to," right? Some kids had probably gotten to that realization, but there wasn't an opportunity to be pushed on that and say, "Who can show me up here on the first try that they can do three rings in seven moves?" How they keep doing that is an opportunity for them to then articulate –

EZIO: Right, right.

PATTY: "Well, I know that this is my placeholder for the first [ring]," or whatever, right? It's a small enough number, you can start to unpack it. They unpack it for themselves. Everybody already had reached that stage.

VERONICA: That first step.

PATTY: Then it's a matter of, "Okay, can we extend the thinking now to four rings?"

EZIO: Right.

In describing the scaffold, Patty rehearsed an alternative lesson format. She built on our sense that, when Ezio introduced the cheat, most students felt confident that they could solved the three-ring puzzle in seven moves. Instead of circulating to groups and asking if they were confident in their answer, Patty proposed a whole-class discussion, inviting a demonstration of this pattern with the prompt, "Who can show me up here on the first try that they can do three rings in seven moves?" In this rehearsal, Patty's imagined dialogue

offered something for students to unpack together, thus scaffolding their articulation of the pattern.

Veronica asked why Patty proposed interrupting the class at three rings –
and not, for example, four:

VERONICA: Patty, you're saying that four rings is still an experiment because
they hadn't formalized their conclusion to three rings?
PATTY: Well, and the advantage of three rings is you get more kids, right?
Some I don't think had gotten to 15 yet.
LANI: It's easy when you're doing four rings to not do the most efficient path.
EZIO: Right.

Patty clarified that three rings was a more inclusive example, because most
groups had managed to generate the minimum number of moves. Ezio seemed
to appreciate Patty's suggestion, further articulating the concern that led him
to the cheat:

> That was my fear, they would waste so much time on the counting, just
> the number. I think that's where most of the time was wasted. We never
> got in-depth with anything else. If I just focused in on the three-ring
> puzzle, … okay.

With this explanation, Ezio agreed that Patty's suggestion would have helped
students move beyond the time-consuming counting phase. He narrated this
as "wasting time," underscoring his goal to go "in-depth" with the task and
the underlying mathematics. Shortly after this, he lamented that telling students the cheat inadvertently "robbed" them of an opportunity to explore the
pattern themselves.

Shifting to some positive observations about the lesson, we pointed out
that students were highly engaged and animated – even if Ezio "robbed"
them of some sensemaking, students were still working hard on the puzzle.
We returned to a clip of Ezio interacting with a group around the cheat and
the recursiveness of the function. He interrupted the conversation to share a
realization:

EZIO: This is where I feel like I totally messed up, that little cheat thing …
That little strategy, I wanted them to […] It's a strategy that can be used
a lot. I think they were so concentrated on the problem that they didn't
– again, I shouldn't 've done it. I should've had them come up with that
strategy on their own, but I wanted them to be aware of that strategy.
PATTY: Yeah, because after you gave the cheat, you actually went around to
all the groups and you essentially asked them all, "Show me that you can
do the cheat."

EZIO: Yeah, yeah, but again, I would totally do what you said, just focusing on the third ring... That was the one thing that was bugging me over the weekend, that I did the cheat. I was in such a rush to just finish up.

PATTY: You wanted them to get closer to [finding the rule].

EZIO: That was just... Yeah. We did a proof by induction for this Tower of Hanoi thing and that was the key, that little cheat thing, to proving the whole thing and I told them instead of like –

PATTY: – Scaffolding it.

EZIO: Yeah

In this exchange, Ezio raised two important insights about using problem-solving heuristics. The first was his acknowledgment that solving a simpler problem is a strategy that "can be used a lot" and he wanted students to be familiar with it. Making problem-solving heuristics visible and learnable bolsters students' perseverance by supporting their capacity to make sense of mathematics. Second, he underscored how he told students the strategy as a *cheat* instead of, as Patty suggested, *scaffolding*, which would have supported students' mathematical agency through collective problem articulation. His insight that he was "robbing" the students of an opportunity for sensemaking was critical for his reconceptualization of this lesson.

Using "Chunking" to Address Multiple Goals

Toward the end of the debrief, we returned to discuss Ezio's goals for the task. Veronica reminded Ezio of one of the mathematical goals he had identified: He wanted to expose students to a non-linear function to contrast it to the linear functions the class had been studying. Reflecting on his lesson design in relation to this goal, Veronica suggested "chunking," or breaking the task into distinct segments – perhaps focusing on finding efficient solutions one day, and the general rule another day. This would allow Ezio to home in on different goals over multiple lessons, using the Tower of Hanoi as an anchor. This notion appealed to Ezio, as he summarized at the end of the debrief:

EZIO: I guess the main takeaway, at least, I feel like I crammed in too much for one day, for one class period. This could've been spread out over a couple [of days]. Just the idea of chunking it – I totally wish I had done that. There were so many things I wanted out of this.

LANI: It's a rich problem.

EZIO: Yeah, it is a rich problem. I just feel like there were a lot of missed opportunities. If I had chunked it and just spread it out, I would've been able to achieve most of those goals, I think... Yeah.

By "chunking" the task, Ezio could slow it down and elicit more of the mathematical richness of the puzzle. This would, in turn, support his goals for the task: challenging students, helping them develop perseverance, exposing them to various functions, and using a scaffold to support their problem-solving skills.

What Made This a Critical Moment: Leveraging Rich Tasks for Depth and Agency

Ezio used a complex task to encourage his students' perseverance through difficult problems. He posed a challenging question ("How do you know that [your number] is the minimum?") but, to accelerate student progress, he used the "cheat." This bypassed an opportunity to support students' perseverance and their articulation of their mathematical sensemaking. Patty's suggestion of the scaffold offered an alternative to help students make progress on the solution while supporting their mathematical agency. In this way, the scaffold aligned with Ezio's pedagogical responsibility to helping students persevere through challenging problems.

Ezio identified chunking as another key takeaway: Taking the time to engage deeply with a mathematically rich task would allow for increased opportunities for students to construct mathematical meaning. The understandings embedded in these practices reflect an important conceptual shift toward ambitious and equitable mathematics instruction: namely, designing instructional activities to prioritize sensemaking over covering material.

Lee Bellver: Supporting Conceptual Understanding

Background and Context

At the time of our study, Lee Bellver had taught for about 15 years at Falconer Middle School. As department chair, Lee had hired most of the math department, including Doha Arzoomanian. Lee had deep affection for his students, which was evident in his classroom. When we observed him, he often joked with students and asked about their lives and their interests. In interviews, he consistently described his goals in terms of students' future success and well-being. As a middle-school teacher, he wanted to ensure that his students were confident in their mathematical abilities and that they were well-positioned for success in high school, college, and their future careers.

Lee's Inquiry: Reducing Students' Teacher-Dependence Through Stronger Student Discourse

Throughout our study, Lee raised questions about supporting students to rely less on him as the teacher and to instead see themselves as generators of mathematical ideas. He was strongly committed to cultivating students' positive mathematical identities. To support these goals, Lee was working on improving his questioning strategies so that he could prompt students to develop mathematical ideas for themselves and not wait for his explanation or direct guidance.

Where Lee's Learning Started

At the beginning of Project SIGMa, Lee knew the importance of rich tasks for fostering students' mathematical sensemaking, and he was working on developing structures and routines to support ambitious and equitable instruction. Lee's co-inquiry questions consistently centered on issues of facilitation, with an aim toward cultivating student understanding. In Falconer VFF 1, for instance, he asked, "How can I be more of a facilitator and less of a direct instructor?" In Falconer VFF 2, his question was, "How can I improve my whole-class questioning?" Later, he asked about implementing specific questioning strategies that he learned at a PDO workshop. Lee consistently worked on improving his facilitation of students' sensemaking.

Lee's Critical Moment in Falconer VFF 10: Using Resources to Anchor Students' Sensemaking

In Falconer VFF 10, Lee had an interaction with two students that did not go as he had hoped. During the interaction, Lee felt unsuccessful in helping them understand the problem they were working on; he described being most frustrated that he was not able to "spark" their sensemaking. At the start of the debrief, he was still thinking about this moment, trying to figure out what he could have done differently. The interaction became the focus of the VFF debrief; even though it was not something Lee had articulated beforehand, it was clearly weighing on him. As they analyzed the interaction, Sammie, Grace, and Doha helped Lee reconceptualize his instructional choices and refine his pedagogical judgment.

Lesson Overview

In this 8th-grade Algebra 1 lesson, Lee gave students three sets of questions over various review topics, which he likened to courses at a restaurant: appetizers (easier problems, typically requiring calculations), main courses (slightly challenging problems that applied mathematical concepts to real-world situations), and desserts (more challenging problems that required creative thinking). Working in pairs, students were expected to complete the "appetizer" problems and choose at least one "main course" and one "dessert." Students had already learned the mathematical content needed for the problem sets; they were reviewing in preparation for end-of-year tests. Accordingly, Lee aimed to do little direct instruction. Instead, he intended for students to use the resources at their disposal – including prompts that he had embedded in each problem, and other materials around the room – and collaborate to solve the problems. When students asked for his help, Lee planned to ask questions to help them make sense of the problem, rather than tell them what to do.

During class, Lee approached two students – Elias and Sergio – who were working on a question that invited conceptual thinking about volume. They were given dimensions for wooden matchsticks and a tree. They were told to approximate the tree's volume as a cone – a very tall, elongated cone – and the matchsticks' volume as rectangular prisms. They were then to approximate the number of matchsticks that could be made from the tree.

Elias and Sergio had correctly determined the volume of the tree and one matchstick, but it was unclear if they understood the context – specifically, how a tree trunk resembles a cone, and how that cone could be cut into matchsticks. Lee spent about eight minutes asking them questions to help them figure out how to relate these two quantities, but the conversation stalled. Eventually, Lee realized that he needed to check in with other groups. He was frustrated with himself for not coming up with more helpful questions.

A Challenging Year

At the start of the debrief, Lee described several personal and professional challenges that he had been facing. Throughout the year, Lee had been out on family leave and had other absences. As a result, he felt like he had not been as present in the classroom as he would have liked, and he saw a negative impact on his students' engagement and learning. He framed his personal circumstances as creating a conflict with his pedagogical responsibility to be a consistent and loving presence in his students' lives:

> Love comes from shared experiences, not shared values – I really believe that – and I think I've missed out on a lot of shared experiences this year. And because of that, I know I'm trying good things but they're not working as well as I hoped they would.

Lee went on to describe how his absences impaired his ability to build relationships with students. Ultimately, Lee said, "I do not feel really good about my teaching self right now." But, he added, he was grateful to have an opportunity to have "extra time to think about all this" in the VFF debrief.

In addition to feeling disappointed in himself, Lee described institutional pressures that made the year even more difficult. He and Doha were facing pressure to raise students' test scores; they prepared additional materials and devoted class time to helping students prepare for these high-stakes assessments. For many of their 8th-graders, Lee and Doha were also preparing them to take another end-of-year assessment that could grant them high-school Algebra credit. Lee said that he was "obsessed" with helping students pass Algebra. He and Doha did not necessarily feel like the high-stakes tests were accurate reflections of students' learning, but they recognized that high scores could open important opportunities for high school and college.

"I was Trying so Hard and not Getting Anywhere"

After sharing these challenges, Lee asked to revisit his interaction with Elias and Sergio, saying that, "I was trying so hard, and I felt like I wasn't getting anywhere." He continued to elaborate on his frustration:

> As it was happening, I was thinking, "Man, I cannot." But you can see on the camera my face was frustrated, and I wasn't frustrated with *them*, I was frustrated with *myself* for not – I just could not get to where I needed to get to as a teacher in my head. Then to formulate the question that I felt they needed.

At the start of the clip, Lee pointed to Elias and Sergio's calculation of the volume of the match – which they correctly found to be 1/50 in^3 – and asked, "So that's going to represent your volume of what?" Lee wanted them to recognize that it was the volume of one matchstick, but they did not seem to know what Lee was looking for. They guessed things like "the cone" and "the width." Unsatisfied with their answers, Lee pivoted to ask about the volume of the tree, which they had correctly calculated as 83.27 ft^3:

LEE: Let me ask you a different question, let's look at this work, okay? What does 83.27 represent?
ELIAS: The cubic feet.
LEE: The cubic feet of what? The volume of what?
SERGIO: The height.
LEE: So height is a dimension. So, when I say what thing, I mean the volume of the building or the volume of the bowl or the cube.
ELIAS: The cone.
LEE: So this problem has to do with real life stuff, right? What are the two things that we are finding the volume of in this problem?
ELIAS: The cone is the radius.

Lee interrupted the clip to comment on Sergio's answer, "The cone is the radius":

> So, he was like, "the cone is the radius," right? Then I was like, "I don't even want to get into this with you because I just want to" – you know what I'm saying? So, when do you make those choices? I was like,

"I don't want to talk to you for 10 minutes about the fact that a cone has a base, which is a circle, and that circle has a radius, because right now I just really want you to realize that the cone is what they are using as a model for the tree trunk." Right? But like I never – I don't know, because there's just so many mis – I'm not saying misconceptions. Because I am not even going to assume that when students say things like that, that it's a misconception. Sometimes it's just a lack of words, right? Or practice with those words or whatever. But yeah, I don't know. When do you – how do you decide? Plus, I'm already – you've seen this – I mean this is like three minutes, right?

In this turn, Lee clarified his dilemma: Should he have followed up on Elias's apparent confusion between a cone and a radius? Or should he have moved on, since he had already spent significant time with these students (and there were many other students to check on)? Lee also wondered how to interpret the students' confusion: Did they have a misunderstanding about the mathematical content? Were they unfamiliar with the vocabulary? Or were they having difficulty expressing their ideas? Certainly, there were no obvious answers to Lee's quandaries.

Doha commiserated with Lee, saying that she has felt like she cannot "do the mini-lesson at that point, when I have another 42 [students]" in her classroom. She also pointed out that they had not yet talked about circles or cones with their students that year. Even though these concepts hand come up previously, in 7th grade, Doha said "it depends on how much background knowledge they [the students] have," suggesting that they may not remember enough 7th-grade content to make headway on the problem.

Building on Doha's comment about background knowledge, Lee suggested that he could embed more resources – like an image of a cone – into the task. We agreed that an image might have helped Elias and Sergio visualize what was happening in the problem. Sammie noted that, with a visual representation, the students might have recognized that "the cone is the radius" was incorrect.

In the next segment of the clip, Lee asked Elias and Sergio more questions that they did not answer confidently. After Elias said that "the cone is the radius," Lee replied:

LEE: But what does the cone represent in real life for this problem? Sergio, read me the first part of the problem.

SERGIO: [*Reading from the problem*] "Matchsticks are rectangular prisms of wood measuring approximately 1/10 inch by 1/10 inch by 2 inches."

LEE: So, you use those numbers, right? To get 1/50th. So, what does 1/50th represent?

SERGIO: The rectangular prism.

LEE: Yeah. And what are the rectangular prisms in reality for this problem?

ELIAS: The tree trunks, right?

LEE: Not the tree trunks, the what?

ELIAS: The width?

LEE: You're telling me a dimension, but I'm asking for a thing, like an object. What object?

SERGIO: The cone?

LEE: You think so? So, where did you get all three of these numbers?

Doha interrupted the clip, asserting, "They don't get the idea of the problem." Lee agreed, saying that they did not have any "anchors" for the problem – they needed a grounding idea or image for what the problem was referring to. In this case, Lee said, an anchor could be "just a picture of a cone, a picture of a tree." Using a visual representation to illustrate how a tree trunk is similar to a cone – with a wide, round base that tapers to a point – might have helped Sergio and Elias understand "the idea of the problem."

We continued watching the clip – which showed Lee asking more questions that seemed to further confuse the students – and paused intermittently to diagnose the interaction. Lee was concerned that the students did not realize that their calculation of $1/50$ in^3 was the volume of one matchstick – though, he conceded, "There's obviously a lot of other things that needed to be addressed."

As the conversation unfolded, we considered many potential points of confusion and ways that Lee could have responded: He could have found images of cones and tree trunks to give them visual anchors for the problem. He could have shown them physical matchsticks and a cone and asked them to estimate how many matchsticks could fit in the cone – and then the number of matchsticks that could fit inside a tree. He could have asked questions to see if they understood the concept of volume, and how that related to the problem context. He could have helped them make sense of the context itself – since wooden matchsticks might have been unfamiliar, and they might not have understood how tree trunks could be approximated as cones.

As we considered these alternative pedagogical actions, Lee recognized that Elias and Sergio were trying to make sense of the problem – and that they had, indeed, calculated the volume of a matchstick and the volume of the cone. He repeatedly emphasized how hard they were working, saying, "They are interested – they want to figure it out." He framed the interaction as a failure on his part, saying that he "screwed up" in asking questions that were not helpful and that he should have supported them better, to "have them be able to struggle more productively."

Ultimately, Lee realized that the resources he had embedded in the problem were primarily procedural. He had anticipated and provided supports for procedural issues that students might encounter, like calculating the volume of a cone. By watching and discussing the video clip, he recognized that he needed "to think of not just procedural roadblocks, but conceptual roadblocks" and provide resources to help students understand a task's underlying concepts and connect them to the context.

Lee seemed energized by this realization. He described how he could prompt students to make a sketch of the problem and how he could help them visualize what the question was asking. As we ended the debrief, with Lee's mental gears turning, he said, "I got a lot out of this. I know we didn't watch a ton of video, but I definitely have a bunch of things that I can apply."

What Made This a Critical Moment: Gaining a New Perspective

In this VFF debrief, Lee was able to re-watch a frustrating interaction with fresh eyes and with support from Doha, Sammie, and Grace. This allowed him to get different perspectives on what had transpired, re-imagining how it could have gone differently (or how it might go differently in the future) – the learning opportunities we tried to foster through our facilitation. At some points, the VFF debrief felt like a grim autopsy; Lee even joked that it was "punishment." But as we watched Lee's conversation with Elias and Sergio, he was able to articulate what he had been thinking in the moment – including the dilemmas he faced – alongside our interpretations of what happened.

By bringing these resources together, this VFF debrief allowed for collective interpretation of a frustrating teaching moment at the end of a personally challenging year. Lee came to a deeper, more integrated understanding of anticipating students' misconceptions or "roadblocks" and better supporting their understanding as they work on challenging tasks. In his Exit Interview, Lee explained that he could not have come to this understanding by reflecting or watching the video on his own:

LEE: When you're looking at it the next day, it's easier to take a deep breath and be like, "Oh man, maybe if I hit them with this question." Or like, on the last debrief, there was a big chunk of time I spent with these two kids [Sergio and Elias], and I think it was like a really big takeaway, which was like – I don't think they had anchors to the problem. When I say anchors, I mean they couldn't *picture* what a tree looked like, and they couldn't picture what a [cone] looked like, and they couldn't see how those two things relate to each other, and how important providing all those resources are to the kids. And I think in that moment, I was really patient and kind, but I spent like, ten minutes with those kids. At the end of it, they still didn't have any sense of clarity, but it was cool to see it, right? Because without that, I don't think I would have gotten to that anchor piece, without watching it the next day.

LANI: You left the lesson knowing, "I spent like an overlong amount of time compared to other groups and I don't feel like they got what they needed."

LEE: Definitely not.

LANI: But you didn't know what they needed.

LEE: Exactly. I couldn't have gone home that afternoon, and thought about it, and been like, "Oh, I bet they just have no idea what," it took Sammie and Grace to be like, "Do you think they knew?" Like capturing it in a, "Do you think, do you think they know what a cone is like?" Well, I made an assumption that they did, but it's not – in that case, it was not an accurate assumption.

Unquestionably, Lee was a reflective teacher – at times, he even went back to watch video from VFFs to do another round of analysis on his teaching. Despite that, in this instance, he felt that his reflective inclinations would not have adequately supported his learning about that interaction. The hybrid space of the VFF debrief, however – bringing together rich representations of practice and multiple perspectives on Lee's practice – allowed him to come to new understandings of the limitations of his practice (i.e., that he was providing procedural, but not conceptual, supports) and to generate new ways of supporting his students in the future by ensuring that he provide anchors to support his students with challenging tasks.

Discussion

The rich representations of practice that we used in VFF cycles allowed teachers to look back at their instruction in great detail. In addition, our facilitation – with its emphasis on collective interpretation and refining teachers' pedagogical judgment and problem framings – prompted us to listen to them in particular ways during our debrief conversations. As a result, each debrief took a different shape. We designed the VFF process to respond to teachers' needs – building on their strengths and centering their questions and problems of practice (see Chapter 5). This approach allowed us to support teachers in their learning trajectories, helping them bring their pedagogical actions in line with their commitments and responsibilities to their students. For Doha, Ezio, and Lee, this created critical events to support their understanding about key aspects of their work.

For Doha, her commitment to fostering student ownership of mathematics conflicted with her practice of interrupting her confused students with mini-lectures intended to keep them on track. The Falconer VFF 6 debrief, with its emphasis on imagining alternative pedagogical strategies, offered her opportunities to learn practices that addressed the dilemmas she faced. She incorporated one new practice, the huddle, so completely that she adapted it further to address the ongoing challenges of her large classes.

For Ezio, telling students how to solve a problem conflicted with his desire to help students learn how to persevere through challenging problems. He undermined students' perseverance by offering them the cheat. During the debrief, this contradiction became evident, and he imagined an alternative of using a scaffold – *solving a simpler problem first*. This would offer students an approach that could be used in many problem-solving situations, giving them greater mathematical agency. Veronica's contributions also supported his learning when she suggested chunking the lesson into segments, another alternative that would relieve the pressures he felt.

For Lee, his commitment to supporting students' sensemaking went unfulfilled in a frustrating moment, where he felt ineffective at helping Elias and Sergio make sense of a problem. Listening carefully to what the students did (and did not) understand helped to clarify the missing element to his use of challenging tasks: the need for conceptual anchors. With the team's support, Lee identified this as an important oversight in his planning, which gave him new ways to support students' conceptual understanding.

Looking across these cases, we note important commonalities. First, the learning events unfolded in similar ways. They began with unsatisfying events – Doha's mini-lectures, Ezio's cheat, and Lee's frustrating exchange. The VFF resources allowed the focal teachers to review these moments with colleagues and with the research team. This, in turn, allowed for clearer diagnoses of what led to the unsatisfying events – Doha's dilemma about balancing small-group and whole-class discussions, Ezio's realization that students' thinking was not surfaced, and Lee's insight about the underlying conceptual confusion. After establishing a common understanding, we brainstormed alternative pedagogical actions that could better support students' sensemaking. For Doha, this was the huddle; for Ezio, the scaffold and chunking; and for Lee, conceptual anchors. These just-in-time resources could not have always been anticipated, but they addressed teachers' immediate learning needs and thus had tremendous relevance and cognitive salience for their work.

Second, there were common themes across these debriefs. Doha, Ezio, and Lee all struggled with a common problem in ambitious and equitable mathematics instruction: how to interpret and respond to students' difficulties once they have started a task. For Doha, her impulse was to interrupt the class and offer mini-lectures to get students back on track. Ezio wanted to accelerate students through the lesson by offering the cheat. Lee tried to talk through Elias and Sergio's confusion, even though it meant spending a disproportionate amount of time with them when other students also needed his attention. By revisiting those moments in the VFF debriefs, the teachers were able to articulate the tensions they were managing and imagine alternative ways of responding in the future.

Notes

1 In Year 2, Ezio chose not to continue with our study, in part because Veronica left Rees to teach at Banneker High School. This meant that Ezio did not have a PDO colleague (or another close colleague) to bring into the VFFs, and he faced additional work-related pressures and time constraints.

2 *Markers* refer to the audio recording devices we placed on student tables to capture student-to-student conversations.

References

Schoenfeld, A. H., Baldinger, E., Disston, J., Donovan, S., Dosalmas, A., Driskill, M., Fink, H., Foster, D., Haumersen, R., Lewis, C., Louie, N., Mertens, A., Murray, E., Narasimhan, L., Ortega, C., Reed, M., Ruiz, S., Sayavedra, A., Sola, T., Tran, K., Weltman, A., Wilson, D., & Zarkh, A. (2020). Learning with and from TRU: Teacher educators and the teaching for robust understanding framework. In K. Beswick, & O. Chapman, (Eds.) *International Handbook of Mathematics Teacher Education: Volume 4* (pp. 271–304). Brill Sense.

8

LEARNING ABOUT TEACHING OVER TIME

With Nadav Ehrenfeld and Elizabeth Metts

At a basic level, questions of learning are questions of how people's under-standings and practices change over time. Sometimes, as in the last chapter, those timeframes are short. Brief experiences can provoke moments of insight that increase the cognitive salience of certain teaching practices, making new understandings and approaches relevant and worth adopting. In this chapter, we move beyond the epiphanic and look at how our co-inquiry activity sup-ported teachers' learning over longer time periods. Using the cases of Brad Miller and Julie Woodman, we show how teachers developed important prac-tices and concepts about ambitious and equitable mathematics instruction through repeated experiences with the VFFs.

Our underlying assumption is that many practices of ambitious and equi-table teaching are difficult to adopt, especially if they differ substantially from teachers' previous experiences or run counter to the norms of schooling. For this reason, teachers need opportunities to consider new practices in light of their existing repertoires to make them meaningful in their classrooms. As teachers adopt new instructional perspectives, they must figure out what works for them and their students – how different approaches support their goals – and reconceptualize core aspects of teaching in the process.

The two cases in this chapter illustrate teachers' learning about different pedagogical practices. For Brad, a moment of insight in a VFF led him to reconsider student collaboration, making modifications to his groupwork practices over two years. Julie's learning converged around the connections between instructional design and student engagement, which gave her new ways to make her classroom more inclusive. Interesting contrasts surface as we consider their cases: Brad's learning trajectory focused on refining a set

DOI: 10.4324/9781003182214-10

TABLE 8.1 Overview of Brad's and Julie's learning.

	Brad Miller	*Julie Woodman*
Where they started	An accomplished teacher and learner with strong student rapport who used a variety of discourse structures	A teacher deeply invested in caring for students and providing authentic, inclusive learning experiences
Focus for their learning	Student groupwork	Student status
How their learning built over time	From keeping students on task using participation techniques to equitable participation and seeing how students take up his feedback to supporting students to see each other as resources	Recognizing additional connections between instructional design and student engagement and consequently, recasting student learning dilemmas as lesson design dilemmas

of related practices around groupwork, whereas Julie's trajectory was more exploratory, as she sharpened her understanding of the relationship between instructional design and students' mathematical engagement. Brad's case is easier to see through common lenses of teacher learning, while Julie's is more abstract – although no less crucial. Importantly, our perspective on teacher learning as situative concept development captures both trajectories.

Brad's and Julie's cases, summarized in Table 8.1, are presented chronologically to capture the shifts in the teachers' practices and understandings over time. We end the chapter with a discussion of what these cases reveal about teacher learning.

Brad Miller: Learning About Groupwork Facilitation

The VFFs supported Brad Miller's learning about facilitating groupwork more inclusively and intentionally. In our early visits to Brad's classroom, we recognized him as an accomplished teacher who had good rapport with students and developed clear, well-organized lessons focused on important mathematical ideas. To support students' mathematical talk, he used various discourse structures, including think-pair-share and notice-and-wonder (Ray-Riek, 2013), which facilitated small-group exploration followed by extensive whole-class discussions.

Our initial observations showed that, in conversations with small groups, Brad did not attend to who was included in these discussions – an important component of ambitious and equitable mathematics instruction. For example, when students called Brad to their tables, he typically discussed questions with the individual who called for him, without involving their groupmates or prompting students to turn to each other as resources. Brad quickly answered students' questions, but often without eliciting their thinking; this limited students' sensemaking opportunities. During Project SIGMa, Brad shifted his facilitation practices to better support students' collaboration.

Background and Context: Commitments to Students, Colleagues, and His Own Professional Growth

Brad had taught at Noether High School for approximately a decade. He was passionate about teaching, mathematics, and technology; he was also deeply committed to his and his colleagues' professional growth. Brad described teaching as "a never-ending job. It's a tireless job that you can never be satisfied with what you're doing, and you should always try to be getting better" (*Interview, February 2018*). It was not surprising, then, that he was among the first volunteers to participate in VFFs. Brad's parents were both teachers, and he told us that, after working as a sports coach, camp counselor, and tutor, teaching seemed like the right career for him. At Noether, he taught Algebra 1, AP Statistics, and Geometry; he also led the school's Algebra 1 Professional Learning Community (PLC). Brad had tripled enrollment in his AP Statistics course while maintaining a high pass rate on the exam, adding to his reputation as a successful teacher. At the end of Year 1, he was appointed department chair, signaling the respect he had from both colleagues and administrators.

Brad's Pedagogical Responsibility: Commitment to Students' Joy, Achievement, and Mathematical Thinking

As a teacher, Brad wanted students to enjoy math and feel confident in their abilities to solve complex problems. These commitments drove his professional learning. As he explained:

> I'm looking at things that do take more time and do take more effort for me as a teacher, but ultimately are meant to help all students succeed, and that's what I'm trying to do now – to try to look for not just the students who basically would succeed in any classroom regardless of the teacher, but the kids that I can help that maybe would not succeed in a traditional math class.
>
> *(Interview, February 2018)*

One of Brad's priorities was involving more students in mathematics – especially those who are traditionally excluded from high-level courses. In AP classes, he aimed to maintain a high pass rate on end-of-course exams. In statistics classes, his commitment was to help students "understand the world better and be more informed citizens when making decisions" (*Interview, September 2017*). From our first visits at Noether, we got the impression that

Brad's classroom was a welcoming environment for learning interesting mathematics.

To summarize, at the start of Project SIGMa, Brad was a knowledgeable teacher who was passionate about mathematics and committed to his own learning. Given his core commitments – his pedagogical responsibility – teaching strategies that would help him reach students who might otherwise not succeed in mathematics were especially salient. He was eager to find such strategies and sought them out when he attended professional conferences and PDO meetings.

Learning to Learn in the VFFs

Like Abigail, Brad was immediately at ease with the VFF process. During the Exit Interview, when we were discussing the vulnerability others had experienced, Brad conceded that:

> I noticed colleagues being more nervous. I felt comfortable, to be honest. [...] I don't feel uncomfortable with people watching me. [...] I think the ultimate goal is for us to improve, and if we aren't watching ourselves and critiquing, we will just stay stagnant.
>
> *(Exit Interview, May 2019)*

For a combination of reasons – perhaps his focus on constant professional growth or his experience playing and coaching sports – Brad was eager and receptive to learning in the VFFs.

Brad's Participation in VFFs

During Years 1 and 2, we conducted a total of six VFFs at Noether; Brad was the focal teacher in Noether VFFs 1, 3, and 6 (Table 8.2). We returned in Year 3 for a Member Check visit to see which changes in practice were sustained over time. The VFFs offered Brad a space to (1) sharpen his commitments to promoting student engagement in mathematical thinking; (2) support those commitments by re-designing his classroom around groupwork; (3) incorporate new groupwork monitoring routines; (4) become fluent with these routines; and (5) consequently, develop more complex pedagogical judgment around them.

Trajectory of Brad's Questions

Brad's questions clearly built on one another over time. In Noether VFF 1, Brad asked us to look at his questioning strategies and how he supported students to develop mathematical arguments. In Noether VFF 3, Brad asked to

TABLE 8.2 VFF visits to Noether High School during Years 1–3.

Year	VFF	Date	Focal Teacher	Inquiry Question/Topic	Class
Year 1		Sep 2017	All teachers	Introductions and consents	
	Noether VFF 1	**Dec 2017**	**Brad**	**Questioning and helping students construct arguments**	**Algebra 1**
	Noether VFF 2	Feb 2018	Abigail	Facilitating conceptual student-student talk	Algebra 1
	Noether VFF 3	**May 2018**	**Brad**	**Teacher's feedback and how groups use it**	**Algebra 1**
Year 2		Sep 2018	All teachers	Introductions and consents	
	Noether VFF 4	Dec 2018	Marisa	Does math come from the students or teacher?	Geometry
	Noether VFF 5	Mar 2019	Greg	How does technology support (or not) student collaboration?	Computer Science
	Noether VFF 6	**May 2019**	**Brad**	**Teacher's feedback and how groups use it** **What happens if he doesn't answer students' questions?**	**Statistics**
Year 3		Feb 2020	All teachers	Member check visit	All classrooms

continue this line of inquiry, but with a focus on the feedback that he gave to groups and how students used his feedback. Similarly, in Noether VFF 6, Brad asked us to extend our scope to consider what happened when he left groups without giving them an answer – a practice that he had adopted between Noether VFFs 3 and 6. This trajectory of inquiry showed how he refined his understandings of and practices for groupwork monitoring, with the goal of fostering student interdependence.

Noether VFF 1: Involving All Students in the Group

Overview of Lesson

For Noether VFF 1, we recorded Brad's Algebra 1 class. In this lesson, students were making connections across linear equations (given in point-slope form and slope-intercept form) and their graphs. They worked in pairs on a notice-and-wonder activity: Given graphs with intersecting lines, students were asked to record their observations under the prompts *What do you notice?* and *What do you wonder?* As students worked, Brad circulated around the classroom, saying things like, "Remember, no wrong answers... Just write as many things as you can..." He occasionally checked in with students as he circulated. Brad then asked students to share their "noticings and wonderings" with the class. When he had accumulated a list of student-generated questions, he asked students to work in pairs to answer one of the questions. At the end of the lesson, students shared their explorations in a whole-class discussion.

Start of the Debrief

Lani, Patty, and Nadav facilitated the debrief with Brad and his colleagues, Abigail Graham, Greg Kahae, and Marisa Dawson. At the start of the debrief, Brad framed the co-inquiry around his questioning strategies and how he supported students' argumentation. Lani asked Brad to describe his vision for student participation. Brad emphasized the role of students' talk as "leading to good mathematical thinking, logic, and reasoning." For Brad, discourse was a way for students to develop important mathematical practices.

We began with two short clips. Clip 1 involved a student explaining his thinking to the whole class and realizing an error as he spoke ("Oh snap. Let me reverse that. I was wrong"). We noted Brad's encouraging response and how it affirmed the student's real-time revision. In Clip 2, two students talked about "arguing," and, after watching the clip, we discussed what strong collaborative discussions would look like.

Clip 3: Asking Leading Questions

In a third clip, we watched Brad approaching a pair of students as they worked on the notice-and-wonder activity. In this interaction, Brad only talked with one of the two students, and he used leading questions to guide her to the correct answer. After watching the clip, Brad said:

> I think that I kind of led her and told her, "Do you think point-slope would work?" – like a yes-or-no answer. She kind of had her head down most of the time and wasn't really taking a second to think about what information she knew. [...] I guess I was kind of just trying to, "You know you're supposed to use point-slope here, right? Which one would help? Well, we don't know the y-intercept." And then I kind of just said, "Would point-slope work?"

Brad's summary of his questions showed that he noticed that he had pushed the student toward the right answer, but without encouraging her sensemaking. He did not, however, seem to notice that he only talked to one student; her partner was disengaged from the exchange. Lani pointed this out.

LANI: I'm wondering if you could rewind time and think about – You said one thing, like, "I would try to ask who's leading the questions." But I wonder if there are moves that you could incorporate so that the other student was more involved, cause his body language was kind of like, back [*Lani leans back in her chair*].

BRAD: Yeah, yeah. As if he was just gonna let her do it and then copy it.

After Lani noted that the second student was excluded from the conversation, Brad recognized it as a problem.

This spurred a brainstorm of strategies that Brad could use to facilitate more collaboration. Lani suggested that Brad direct students to talk with their partners by explicitly stating that he would not talk to groups during the first few minutes of the task. Marisa, drawing on her experience with Complex Instruction (see Chapter 2), suggested that Brad use group roles – assigning one student the job of calling the teacher over for "group questions" – to encourage students to collaborate before asking for Brad's help. Both strategies would provide students resources to keep working even when they encountered challenges, offering alternatives to asking leading questions.

Summary of Brad's Learning in Noether VFF 1

Toward the end of the debrief, Brad summarized his learning from the conversation, saying that the debrief made him "more conscious of who I'm going to talk to during partner- and groupwork." He described this as a shift from "managing" students and keeping them on-task to being mindful of inclusion, including who participates and whom he interacts with. He also articulated a new question: "Am I leaving students that are not raising their hands or not participating around areas where I walk a lot of the time, am I leaving them out?" Brad's new question, which emerged from the debrief conversation, was more specific than his original co-inquiry question. Instead of asking generally about questioning strategies and student discourse, he was more focused on how his pedagogical actions influenced students' participation and shaped patterns of inclusion and exclusion.

Noether VFF 3: Helping Students See Each Other as Resources

In Noether VFF 3, Brad continued thinking about these issues in his classroom. Specifically, he asked us to help him look at the feedback he provided to groups during the lesson and how students took up that feedback. This reflected Brad's insights from Noether VFF 1, as he continued examining how his pedagogical actions influenced student collaboration.

Overview of Lesson

In this Algebra 1 lesson, students worked on sketching graphs of quadratic functions. Before starting the main activity, Brad reviewed key features of quadratic functions, including that graphs of quadratic functions have an axis of symmetry that goes through the vertex of the parabola. As students began groupwork, Brad told them, "I want to hear you guys talking. Please work together. Be collaborative." In the task, students described and sketched various quadratic equations. At the end, they answered questions to help them generalize their work (e.g., "Describe any similarities and differences between the four functions you just graphed"). At the end of class, Brad facilitated a whole-class discussion about students' work.

Start of the Debrief

Greg, Abigail, and Marisa joined Brad for the debrief, which Lani and Nadav facilitated. We began by asking Brad to describe his lesson. As he summarized the task, he explained that he designed this lesson around groupwork, but that he was still working on supporting fuller student participation; he noted that these shifts were based on feedback from Noether VFF 1. Brad explained that he selected focal student groups (see Chapter 4) that were fairly heterogeneous in achievement. He was interested in looking at the feedback he gave to groups and how they took it up.

Clip 1: Interacting with a Student Group

We spent much of the debrief discussing a clip that showed Brad interacting with students in a way that was typical for this lesson: He approached the group and answered their questions with a closed exit from the interaction – meaning that he explicitly prompted the students' subsequent work. (In contrast, open exits leave the group with something to think about; Ehrenfeld & Horn, 2020.)

The students – Gaia, Calista, Remy, and Danny – called Brad over and described two difficulties they encountered trying to graph a quadratic function. First, their graph extended beyond the 20-by-20 grid provided on the worksheet. Second, the x-coordinate of the vertex was 2.5, making it difficult to plot on the grid. Brad responded to both these questions in a similar manner. He asked some questions to cue them toward the answer and then ended the conversation with a clear directive of what to do next. For example, he ended the discussion about the vertex with the following exchange:

BRAD: Is it possible that the axis of symmetry doesn't fall on a nice, numbered point?
GAIA: No, yeah, it is.

BRAD: Or is it an integer? I should say, better yet… Is it possible that the axis of symmetry is *not* an integer?

GAIA: Yeah, I think it is. I don't know.

DANNY: Okay.

BRAD: Alright, so I want you guys to always be sort of aware that some parabolas will be really nice and line up with beautiful coordinates, but other times they won't. So, where do you think the axis of symmetry actually falls on this one?

GAIA: In the middle.

BRAD: And where would the middle be?

CALISTA: The middle of 2 and 3

BRAD: …which is?

DANNY: So 2.5.

As he wrapped up the discussion with the group, Brad asked Remy, "Are you with us?" Remy gave a half-hearted "Yeah." In the debrief, Brad explained that he did not want to "leave her behind." The clip ended as Brad went to help another group that had gotten his attention.

After watching this clip, Brad noted that he only talked with Gaia, Calista, and Danny. He started to check in with Remy but had not really engaged her in the group's conversation. He also elaborated what we saw in the video, explaining that the students' work indicated that they had been guessing-and-checking to find the graph, rather than sketching it from the key features as he had hoped in his task design.

Nadav filled in some of the student discussion that took place just before the clip started. Specifically, because the line of symmetry was not an integer, Gaia had proposed to the group that it might not exist. Danny disagreed with Gaia, which is why they called Brad over to settle their disagreement. Nadav noted that this had become a pattern in Brad's class:

NADAV: It seems like the norm when they have this contradiction is to go to you to be the judge […] What can we do to take this disagreement between them [about the axis of symmetry] so that they would take it as an object to figure it out between them and not to turn to you?

BRAD: I wonder if there is a better question I could have asked.

We began brainstorming alternative questions that Brad could have asked. Greg suggested, "Why do you think that?" and Marisa offered, "How did you figure that out?" Lani suggested asking about their understanding of the coordinate plane, to elicit – and then build on – what the students already know. To help Brad develop pedagogical reasoning about which of these actions to choose, Nadav asked him to think about the goal of his conversation with the group. If Brad's goal was to support collaboration, for instance, he might ask students to make sense of the problem together, rather than settle their debate himself.

After we shared specific instructional strategies, Marisa connected our discussion to issues of school and classroom culture. She described how Noether students – in her classes, as well as her colleagues' – were frustrated when teachers asked students to figure things out together instead of telling them the answer. She suggested that it might be because their lessons were "going kind of back and forth" between direct instruction and groupwork, and not using groupwork consistently. At her previous school, Marisa had used a problem-based curriculum. She said that it worked well, "because it's in groups from day one. Every day, students are working in groups – that's the culture of the classroom we built up."

By describing her experiences working with a problem-based curriculum, Marisa broadened the conversation from an interactional lens (*What strategies would be the best in this situation?*) to include issues of design (e.g., working in small groups every day) to build a collaborative classroom culture. This points to the cultural change part of developing ambitious and equitable instruction, noting how daily routines give students new habits that support different forms of instruction.

Summary of Brad's Learning in Noether VFF 3

In this VFF, Brad showed progress toward his goal of including more students in mathematics. Although his exchange with the students in Clip 1 primarily focused on Gaia, Calista, and Danny, he attempted to involve Remy in the interaction as well. He recognized that his questions continued to emphasize right answers and that he did not elicit students' thinking about the problem. The team offered numerous ideas about improving his facilitation, including Marisa's description of the potential of daily groupwork to shift classroom culture.

Noether VFF 6: Daily Groupwork and Reconceptualizing Student and Teacher Roles

In Year 2, Brad took up Marisa's suggestion to incorporate groupwork more consistently. In their Geometry course, they co-planned a six-week statistics unit that was designed for daily groupwork. We filmed one of these lessons for Noether VFF 6.

Overview of Lesson

In our final Noether VFF, we filmed a lesson about one- and two-variable quantitative data. Students were asked to measure their shoe lengths and handspan measurements. In groups, they discussed descriptive statistics (e.g., mean, median, range, standard deviation) for each variable and the correlation between the two variables.

In the beginning of the lesson, Brad explained that the student with the biggest handspan was the only one who could ask him questions on behalf of their group. He further prompted, "I need to see you talking to each other. Come with questions as a group." Then, when individual students asked him questions, he consistently redirected them, saying things like:

- "Who is the big handspan person in the group? Ask him."
- "Did you ask your group? I bet someone in your group knows." [*Student, turning to a peer: "I bet you know…"*]
- "I'm telling you, they have more resources than I do."
- "I feel like your handspan is not the largest. Did you guys talk about it? [*Student: No*] Why don't you talk about it, and I'll come back in a few minutes.

These moves showed that Brad was intentionally facilitating student collaboration.

Start of the Debrief

Nadav and Katherine facilitated the debrief with Brad, Marisa, Greg, and a new colleague, Joseph Park. Initially, Brad's co-inquiry question was about his feedback to the groups and how students used it. Right after the lesson, though, he also asked us, "What happens when I leave a group without giving them an answer?" When Brad introduced the lesson at the start of the debrief, he explained:

> I've had these kids in these groups for the last two weeks. They've been in groups of four, even though Marisa and I have barely done group-work. We did it once in a while, if the lesson called for it, but the last two weeks, since we started Statistics, every single day has been group-work, the entire time.

Clip 1: Teacher Directives as a Catalyst for Student Conversation

Attending to Brad's questions about his feedback and how students used it, the first clip showed a student asking Brad about calculating and interpreting the descriptive statistics. Brad realized that the student had not talked about it with her group before calling him over, and he asked the group to discuss the question before getting his input. After Brad left, the students figured out the problem together without his help. Moreover, once they had their answer, they continued to talk to each other as they moved on to the next question.

The debrief team noted that this was an important moment:

KATHERINE: That move you had made kind of started this catalyst of them working together, using each other.

BRAD: Using each other as resources.

MARISA: I think I need to do more of that, like not answering questions. Because I have that problem too, where they're seated in groups but working individually.

Katherine's description of Brad's redirection as a "catalyst" for the group's collaboration made sense to both Brad and Marisa, with Brad connecting it to his larger goal and Marisa recognizing its potential for her own instruction.

Clip 2: Going Beyond the Group Question

The next clip illustrated how Brad approached groups that had discussed their question together before calling him over. The clip started with a group's conversation about a question in the task: "How would we describe the spread [of the data]?" As the group talked, it was clear that they were familiar with the idea of range and standard deviation as two tools for describing the spread of their data. However, they were confused about calculating the standard deviation, and their result did not make sense to them. Brad first asked if they had talked about spread before asking him, and they confirmed they had. Then, Brad immediately responded by reteaching them what range and standard deviation are.

Because the clip included the students' conversation before they sought Brad's help, it was apparent that he was repeating content that they had already discussed. Acknowledging this after viewing this clip, Brad suggested that, to better build on their ideas, he could have instead asked, "What about spread confused you?" Katherine also suggested asking, "What was your conversation about?" Taking the time to elicit the group's thinking would better position Brad to build on their ideas and support their sensemaking.

Extending the Discussion: How Much Scaffolding is Adequate?

Toward the end of the debrief, Nadav summarized some of the interactions he had seen during the video review, beyond the clips we had watched in the debrief. These highlighted the complexity of figuring out in the moment how to interact with a group. Nadav said, "The question I'm kind of posing for you and for everyone is: How do you make these decisions? If you were being intentional about not giving answers, how do you decide when giving some answers is scaffolding?" Brad described how he judged different groups' readiness to work with less direct assistance (e.g., "I was more hands-off with them because I knew that the girl in the red comes to tutoring every day and she has an A. [...] I knew that they had resources in the group that they could get there"). After Brad shared this, Nadav gently raised another issue:

> I guess the one kind of, I think, trap that we have here is this dilemma – and I like it that you kind of are telling how you observed them, and that's why you felt like you need to support them more, because we don't want to fall into the trap of supporting them more because we have lower expectations for them. [Instead], to support them more because we kind of observed *that they need it in the moment*, and not because we decide ahead [of time] that these groups need more support.

Carefully raising the issue of teacher expectation bias, Nadav urged Brad to consider looking at what students were doing and thinking in the moment rather than approaching them with preconceived ideas, inviting another crucial shift toward building inclusive classrooms.

From there, the discussion went beyond introducing new technical solutions for supporting productive interdependence in groupwork to more nuanced sensemaking about responsiveness in teaching and its relationship to Brad's commitments to reach students who have struggled in mathematics in the past.

Summary of Brad's Learning in Noether VFF 6

This VFF showed Brad's uptake of strategies discussed in previous VFFs. He had developed explicit discourse strategies to involve all students in group discussions by only responding to group questions, something he had learned in Noether VFF 1. He also refrained from being overly directive, instead helping students see each other as resources, which we discussed in Noether VFF 3. In addition, he and Marisa had deliberately designed a unit that incorporated daily groupwork to shift their classroom culture and make messages about participation clearer to their students. Unequivocally, we saw changes in his instruction. Nonetheless, each change pointed to new avenues for Brad's continued learning.

Clip 2 from Noether VFF 6 left a strong impression on Brad. In the Member Check visit during Year 3, Lani and Katherine visited Brad's classroom and discussed our observations with him. During this conversation, Lani mentioned seeing Brad approach a group and ask, "What'd you guys talk about so far?" In response, Brad referenced the discussion of Clip 2:

> I would not have said that if it wasn't for watching that video at [Noether VFF 6], where I gave that group the answer, without having heard from them. I swear, that's my number one takeaway from the video

observations is *that* group – remember the standard deviation, range, all that stuff. And I'm just like, "I didn't give them a chance to say what they had talked about, and I just told them the answer!" And so that is – Yeah, no, and that is the biggest thing that you guys' research has changed for me is, have they talked? or if they haven't talked, that's their fault and they *should* talk. And I used to just make sure that person called me over and then I would answer their question kind of, but I still wouldn't necessarily have a debrief on what they talked about.

Undoubtedly, Brad's instruction in Noether VFF 6 already showed significant changes compared to what we had seen in Year 1. In turn, these changes allowed Brad and the team to have more complex conversations about teaching and student participation, fostering richer learning opportunities for the whole team.

Reconceptualizing Groupwork Interactions

Over time, the hybrid space of the VFF supported Brad in both the adoption of new practices and the reconceptualization of what it means to foster inclusive groupwork. Unlike the brief, yet critical, moments presented in Chapter 7, Brad's case illustrates shifts in his understanding over time as he refined his practice to address new issues that arose with each adaptation.

In Noether VFF 1, he recognized the extent to which his interactions with groups implicated his commitment to inclusion. At the end of this debrief, he described the change in his thinking as a shift from a management perspective to an inclusive perspective, with a new awareness about the importance of whom he was speaking to and when. In Noether VFF 3, the adjustments in his instruction were apparent, as he incorporated discourse strategies for interacting with student groups to help them see each other as resources. In reviewing Noether VFF 3 Clip 1, Brad realized he was asking leading questions when students were stuck, and the team's brainstorm of alternatives again offered reconceptualizations of both his teacher role and the students' roles and obligations toward each other. In Noether VFF 6, Brad recognized the importance of making groupwork a routine part of his instruction to facilitate a cultural shift in the classroom. Having incorporated those changes, Noether VFF 6 Clip 1 showed him that his strategies did, in fact, catalyze students' discussions, offering evidence of the usefulness of this change (see the Utility Premise). At the same time, Noether VFF 6 Clip 2 showed that the group question strategy was necessary but not sufficient for fostering student discourse, as Brad recognized that he reiterated ideas that the group had already discussed because he did not elicit their thinking. The durability of this clip in Brad's sensemaking suggests that he was pulled up short by the viewing (Self & Stengel, 2020; also see Chapter 6), experiencing the disruptive shock of unmet expectations and self-understandings that prompted him to, once again, reorganize how he made sense of things. He realized that, in his teacher role, he needed to do less telling and more listening.

Indeed, our Year 3 Member Check observation documented important changes in his teaching. He intervened less often, instead circulating and listening to groups as they worked. He redirected students to ask their groups before answering their questions. Then, before entering student conversations, he asked them to summarize what they had been talking about so as not to make assumptions and to foster students' ownership of the mathematics. Finally, he used group roles to allocate participation. Overall, there was a lot of evidence that he had developed and sustained more strategies, structures, and awareness about setting up and supporting effective groupwork.

Importantly, he experienced this shift as an alignment of his pedagogical actions and pedagogical responsibilities, which we view as supporting this reconceptualization of practice. As he told us in his final interview:

> It was a goal of mine to be not the leader of material and giver of information from the beginning of my career, but I feel like this year, more than ever, is going the best in terms of the kids taking the majority of the learning on their backs, and me just sort of filling in the blanks where needed.
>
> *(Member Check Interview, February 2020)*

Brad's learning came about as he integrated his pedagogical action and responsibility, adjusting details of practice – including the kinds of questions he asked and how much he listened to students – and connecting these details to his broader instructional goals. In doing so, he reconceptualized what it meant to support students during groupwork, a shift that implicated his understandings of his and his students' roles in the classroom. This alignment also resulted in Brad sharing his increased positive identification with his instructional practice. Brad's case is powerful because it shows one teacher's trajectory from a management view of groupwork to a more inclusive view, something that has been a central project in mathematics education.

Julie Woodman: Connecting Instructional Design to Students' Mathematical Engagement

Brad's learning trajectory involved the refinement of a practice over time, as he worked on his groupwork monitoring to better align his pedagogical actions and pedagogical responsibility. Julie Woodman's learning trajectory involved exploring different facets of an important relationship: the connection between instructional design and students' mathematical engagement. Unlike Brad's learning, which was clearly visible over the sequence of lessons he invited us to film, Julie's learning emerged as a theme as she was investigating issues in her teaching. For her Member Check visit at the end of Project SIGMa, she asked us to observe a revised version of a VFF lesson, enabling us to better ascertain her learning.

The relationship between instructional design and students' mathematical engagement varies depending on many conditions, including the mix of students, examples used, ideas featured in class discussions, and where the lesson falls in the unit (e.g., introduction or culmination of ideas). Ultimately, Julie reframed questions of student participation into more productive and actionable questions of instructional design. Julie's learning is thus more amorphous to illustrate as a case, and yet we (and Julie) agreed that this crucial and complex relationship was, in fact, what she learned through her participation in the VFFs.

Although she explored this broad area of teacher knowledge in different ways, much of her learning was anchored in her evolving understanding of *student status*, a driving concern for her. As we described in Chapter 2, status is a sociological term that, in this context, describes students' perceived academic ability and social desirability. This social reality influences students' interactions as they manage status differentials and engage in status-conscious behaviors (Horn, 2012; Sengupta-Irving, 2021).

At the start of Year 1, Julie focused on status as it impacted classroom dynamics and student engagement, but her understanding of how it operated in the classroom was relatively static. In Julie's pedagogical reasoning, status was a fixed attribute, primarily rooted in students' prior academic experiences. Over time, her understanding of status became more nuanced – students' status could be amplified or minimized through instructional design choices. This reconceptualization offered her more ways to address status dynamics. As our narrative shows, Julie's conceptual change stemmed from her increased capacity to reason about connections between her (often conflicting) pedagogical responsibilities and her pedagogical actions, with a focus on instructional design.

Background and Context: Coordinating Relationships, Commitments, and Instructional Practice

At the start of Project SIGMa, Julie had taught at Fermat High School for several years. During our study, she taught Statistics and Algebra courses. As we described in Chapter 6, she was very close with her partner teacher, Lizette McLoughlin. Julie explained, "There's just a really high level of trust between the two of us" (*Exit Interview, June 2019*). Julie also frequently remarked on how much she liked Fermat and her students. During a site visit in Year 0, she described Fermat as a "happy school." This was apparent in her classroom, as well, where she had strong rapport with students.

Julie's Pedagogical Responsibility: Building Relationships and Broadening Access While Preparing Students for a Test

Julie's commitment to authenticity was at the core of her teacher identity. She described building positive and meaningful relationships with students as the most important – and her favorite – aspect of her job. She told us:

> To me, relationships are first – 100%. I could mess up teaching a lesson, but, for example, that girl walking out upset and me knowing that it might affect the [class] the next day, that's not something I'd budge on. I have the conversations right away, and that's really important to me.
> *(Member Check Interview, February 2020)*

In classroom observations, we saw Julie show great care for students, chatting with them about their individual interests and lives. She was very intentional about making sure they felt seen and valued as people. She was invested in being genuine, expressing distaste for "faking it." This commitment to authenticity was apparent in Julie's lesson design as well. Her lessons were clearly organized, with minutes carefully allocated, but she resisted using structures that felt inauthentic or "phony." While she frequently asked students to collaborate in groups, she did not use group roles or other strategies that would dictate who spoke and when.

Julie also had strong commitments to increasing students' access to mathematics. She and Lizette worked to broaden student participation in upper-level courses. After recruiting academically and demographically heterogenous groups of students, they were deeply invested in making their instruction inclusive. They were concerned that students' prior mathematical histories and levels of achievement affected how they participated in class. This was especially salient in Julie's AP Statistics classes, which she described as "very bimodal," referring to the mix of high- and low-achieving students: Some students entered AP Statistics after taking all the other advanced math courses that Fermat offered, whereas others entered after taking Algebra 2. Julie was concerned that the latter group might be less confident than their high-status peers, and she was committed to supporting all her students to see themselves as competent and capable of rigorous mathematics.

At the same time, Julie was concerned about pacing in her AP classes, with end-of-year tests compelling her to move briskly through the curriculum. "I go really freaking fast because I have to," she told us in Fermat VFF 5. Julie felt pressure to "keep up" with the course content, to make connections across different topics, and to review previous material so that students would be well-prepared for the end-of-course exam.

However, Julie's pedagogical responsibilities were often in conflict. On one hand, she wanted to broaden students' access to rich mathematical ideas by cultivating warm and inclusive classes. On the other hand, she had to keep up with the AP curriculum, which meant moving swiftly through topics. At times, this created a dilemma, since emphasizing speed can often exacerbate status dynamics.

Learning to Learn in the VFFs

Julie, like most SIGMa teachers, had to learn *how* to learn in the VFFs. Like Lizette, she described the importance of building relationships with the research team so she could be vulnerable enough to learn with us. When we asked what she would recommend to other teachers embarking on this process, she said, "I know it sounds crazy, but being vulnerable [...] is the ultimate growth" (*Interview, May 2019*). Though that vulnerability was not always comfortable for Julie, she ultimately felt it was worth it. She described video review as crucial for her professional learning, since she was able to investigate her own questions and receive targeted feedback. Moreover, watching herself teach her own students gave her new insight into how she built authentic relationships with her students.

Julie's Participation in VFFs

To illustrate Julie's learning, we focus primarily on Fermat VFFs 1 and 5 where Julie was the focal teacher, as well as our Member Check visit a year after our intervention ended (Table 8.3). For the Member Check, she taught a revised version of the same lesson from Fermat VFF 5, allowing us to closely compare the changes in her teaching. Although her learning about the relationship between instructional design and students' mathematical engagement was certainly not confined to these events, they show three points in time that illustrate how her practice evolved during our project in relation to her changing understanding.

Trajectory of Julie's Questions

Julie's co-inquiry questions often invoked status as related to students' prior achievement. Her classes were usually large – with as many as 40 students – so it was difficult for her to listen to how each group collaborated. Given the "bimodal" nature of her AP Statistics classes, she was interested in investigating status issues and students' dynamics during groupwork. With this as a starting point, Julie's inquiry often shifted during the VFF debriefs. Much like Abigail's evolving VFF questions in Chapter 5, Julie's questions turned toward issues of instructional design as we considered how her pedagogical actions influenced student engagement.

For instance, in Fermat VFF 1, Julie initially asked about students' mathematical understandings and group dynamics, but the conversation ultimately

TABLE 8.3 Visits to Fermat High School during Years 1–3.

Year	VFF	Date	Focal Teacher	Inquiry Question/Topic	Class
Year 1		Sep 2017	All teachers	Introductions and consents	All Classrooms
	Fermat VFF 1	**October 2017**	**Julie**	**Student talk, group dynamics, and mathematical understandings**	**AP Statistics**
	Fermat VFF 2	Feb 2018	Lizette	Small-group participation: Are all students participating?	AP Calculus
	Fermat VFF 3	May 2018	Lizette	Group dynamics and student participation	Algebra 2
Year 2	Fermat VFF 4	Oct 2018	Lizette	Student participation and mathematical identity development	AP Calculus
	Fermat VFF 5	**Feb 2019**	**Julie**	**Status issues and dominating students**	**AP Statistics**
	Fermat VFF 6	**Mar 2019**	**Julie**	**Connections between math and the task and sticking points**	**Algebra 2**
Year 3		Jan 2020	All teachers	Member check visit	All classrooms

focused on how the examples that she provided influenced these issues. In Fermat VFF 5, she asked about status issues and students who dominated groupwork; once again, the conversation evolved to consider how her lesson design and pacing could exacerbate or mitigate status differentials. Ultimately, by shifting the conversation toward instructional design, Julie took up more actionable framings of her problems of practice. Rather than seeing status as a static feature of individual students, Julie learned to design lessons to intentionally mitigate status differentials. We saw this most clearly in our Member Check visit, as we observed a revised version of the Fermat VFF 5 lesson.

Fermat VFF 1: Classroom Design and Student Engagement

Overview of Lesson

For Fermat VFF 1, Julie taught an AP Statistics lesson about non-sampling errors, which result from biased or skewed survey questions and not from an unrepresentative sample. She began the class with a homework review. Then she introduced non-sampling error with an extended definition and several examples. Students compared two versions of questions that sought the same information – one with bias and one without bias. For instance, one biased question was, "Do you *still* watch cartoons?" which was revised to the more neutral question, "Do you watch cartoons?" In small groups, students generated pairs of survey questions, one that would result in response bias, and an alternative, less-biased question to seek similar information. At the end of class, Julie asked each group to share their example. For her VFF co-inquiry question, Julie asked us to pay attention to student talk, small-group dynamics, and students' mathematical understandings.

Start of the Debrief

For the debrief, Brette and Sammie met with Julie and Lizette during their common planning period. We first watched two clips of Julie interacting with groups during the homework review. Hearing students' voices, Julie elaborated details of their personalities. She described one student as getting "feisty sometimes," and another as a "class clown." As we pivoted to the heart of the lesson, Julie recalled that the students generated "intense" questions to illustrate response bias, saying, "They go to the most extreme [topic] that they can think of." Brette confirmed that observation and mentioned that we had, in fact, identified clips where students considered different topics – including some that were "intense." She offered a few debates for Julie to choose from. Julie asked to listen to "the abortion one."

Clip 3: "Intense" Questions

In Clip 3, a group considered various topics to generate an example of non-sampling error. Two students suggested the topic of abortion, invoking uncomfortable laughter as well as lively discussion. When Julie checked in with the group, one student said that they would ask both pregnant and non-pregnant people if they "would ever abort a baby." Julie pointed out that this would illustrate sampling error, not response bias – that is, the students proposed changing the sample (pregnant and non-pregnant people), not the question itself. As they grappled with incorporating response bias, Julie suggested that they try adding additional information to the question, like a biased description or definition. As Julie left to check in with others, the group searched the Internet for more information about abortion. Some students expressed discomfort with the images and headlines that came up in their search.

At the end of class, this group did not share questions about abortion. When Julie called on them, one student said, "Ours is sad. Ours just made me sad. I changed it." The student proposed asking, "What's your least favorite quality of your girlfriend?" with the girlfriend present and not present. While this could illustrate response bias (i.e., people might say different things in front of their girlfriends and on their own), the open-ended nature of the question would make it difficult to use statistics to show how the responses were biased.

After watching the clip, Brette and Julie began to explore the students' questions in light of the mathematical goal:

BRETTE: I thought it was interesting that they came up with that example, which –

JULIE: Right after.

BRETTE: Well, no, like – What's your least favorite quality of your girlfriend in front of her or not? That's – I don't know if you could, if that would give the data that you'd be looking for [*that could be analyzed for its bias*].

JULIE: I mean, well. Yeah, it wouldn't because… Yeah, we need a numerical – or like a yes-or-no question. Yeah, that's true. That wouldn't give the right type of data, but they were just desperately grabbing.

In this exchange, Brette noted the type of data that the students' question would yield – an open-ended categorical list of qualities. Julie recognized this as problematic since response bias is typically shown with numerical or binary categorical data (e.g., yes-or-no). Julie pointed out that the group was "desperately grabbing" to come up with another question, since they did not want to share an example about a "sad" topic.

Clips 4 and 5: Limited Examples

In Clips 4 and 5, we watched other groups debate various topics and generate biased questions. In Clip 4, students asked about college plans: "Do you plan on going to *real* college or community college?" For an unbiased alternative, they considered asking "What kind of college would you like to go to?" but recognized that the answer choices were too broad. They revised this to a question with binary choices: "Do you plan on attending a four-year college or two-year college?"

In Clip 5, a different group – Tommy, Delilah, Hillary, and Claire – brainstormed topics for their questions: food, video games, and weight. Tommy proposed the question, "How much do you actually weigh?" Delilah rejected this, saying, "No, it has to be, 'Do you consider yourself thick?'" Hillary and Claire quickly agreed with Delilah, saying, "Yeah, it has to be yes or no." Eventually, they settled on a moral question: "Do you consider cheating okay?"

After watching these clips, we returned to the issue of data type. Each group generated questions that provided categorical data, usually with binary answers (e.g., yes or no, four-year or two-year college). When Tommy suggested a numerical question ("How much do you actually weigh?"), his groupmates rejected his idea; they were certain that they needed a yes–or–no answer. Julie recognized this as a limitation of her instructional design:

JULIE: They could say they get a percentage of "yes" that they could know the direction that the bias is going. I think that's probably how they thought of it.

BRETTE: Then, excluding anything that would be –

JULIE: – An average or something.

BRETTE: Right, like asking, "how much do you weigh?"

JULIE: Yeah, that's a good one

BRETTE: You would self-report in a downward direction, but you would use an actual scale or anonymously report [*for an un-biased response*].

JULIE: Yeah, because all my examples were "yes-or-no," so that's probably why they didn't think outside the box in that way. Interesting. That would be something to bring up. We get into those differences later in the year between the proportion and the mean – that's a big distinction we make, but we haven't yet. I would maybe add an example next year with averages, numerical averages. They could think of it the same way.

Through this discussion, Julie realized that all the examples she introduced had binary categorical (and not numerical) answers, which unintentionally limited students' responses. She later elaborated on Tommy's suggestion:

> The weight one would have been better if they knew that was a possibility. That was a really good one, but they just kind of thought that it doesn't work. Then it became categorical. [...] They moved right past it.

Hearing her students consider – and reject – a numerical question confirmed the impact of Julie's pedagogical actions: By only offering examples with binary data in her introduction, she unintentionally limited the kinds of questions students considered. This conflicted with Julie's pedagogical responsibility to support students' conceptual understanding of statistics.

Linking Instructional Design to Student Status

As we unpacked students' understandings of non-sampling error, we also considered how students' status influenced their discussions. Lizette had taught some of Julie's students in AP Calculus and knew how they collaborated. She was curious about how groups decided which topics to explore:

LIZETTE: Was it just, like, whoever spoke first?

JULIE: Yeah

LIZETTE: Then once a question came out, did they just kind of stick to that question?

BRETTE: A lot of them came up with other ideas, but then they circled back to the original question

JULIE: Okay, interesting. I think kids probably do have a habit of going with the first one.

LIZETTE: I feel like that happens in my Calculus classes, too. I'm worried about one kid always starting the conversation, being the dominant one. I

wonder if it's because the first person who shared was normally the dominant one in that group and so they circled back to it because they're like,
"Oh Gordon said that, and he's our person."

In this exchange, Lizette connected part of Julie's instructional design –
specifically, unstructured small-group discussions – to status dynamics. We
agreed that this was probably happening across focal groups. Lizette pointed
out that it might be why Tommy's groupmates rejected his suggestions to
ask about weight; she said that in her own class, Delilah had seemed to treat
Tommy as an "annoying kid brother." Julie suggested that he had low social
status within her class too, as he was one of the few 11th-graders in a class with
mostly 12th-graders. Lizette shared another anecdote about a student, Carrie,
who was in a group with two "really quick" classmates. Carrie liked to think
through things on her own, but her groupmates rarely gave her the space to do
so. Lizette relayed that Carrie was frustrated by that dynamic in both Lizette's
and Julie's classes.

With this insight, Julie recognized that aspects of her lesson design were
exacerbating (or at least not addressing) patterns of dominance and exclusion,
which conflicted with her pedagogical responsibility to support inclusive collaboration. To address this, Julie suggested that she could give "individual think
time first – I do that sometimes." Brette agreed that having students consider
a question individually before talking with their groups would give more students a chance to think through questions. Sammie suggested asking everyone
to write down an idea before talking as a group, so that each student would
have something to share. As we discussed various strategies, Julie expressed
some hesitance at imposing more structure into group's conversations:

> I try things and then they last for like… (*voice trails off*). There were some
> methods I was trying this year that lasted for like a week and then I was
> just like, "This is annoying. This isn't my personality." There are things
> that work, but then when you have so many strategies, there's only a cou-
> ple things I focus on, and when I try to add more, they just kind of fall
> off and they're not consistent enough for the kids to take them seriously.

She went on to connect this to her experiences in structured conversations:

> When I was a student, I hated that, because I was the quiet one… Give
> me some time and I would find something to say, but I didn't like when
> people forced me to give input when I wasn't ready. I felt like I was just
> making something up that wasn't really authentic.

While Julie seemed open to trying some structures that jibed with her
personality – like giving individual think time at the start of groupwork – she

was reluctant to try anything that felt inauthentic or that forced students to share before they were ready. At the same time, she recognized that her commitments to authenticity conflicted with her goals of facilitating equitable collaboration.

Summary of Julie's Learning in Fermat VFF 1

In Fermat VFF 1, Julie viewed several video clips showing how students construed her response-bias task as requiring them to use binary categorical (and not numerical) data. As we listened to students' discussions, she re-interpreted students' engagement with the activity. Groups' difficulty in generating examples was not, as she originally thought, simply a consequence of a new topic, but also at least partly related to her set-up of the task, which inadvertently cued them to construct certain types of questions and reject others.

At the same time, Julie continued to reflect on status issues. As we reviewed the video, Julie and Lizette spent a lot of time describing groups' interactions, with Julie offering multiple interpretations of what we saw. When we identified patterns of dominance in one of the groups, we suggested discourse structures for Julie to use to help address these dynamics. However, Julie resisted these suggestions, explaining that such strategies tended to "fall off" in her instruction because they felt artificial. Despite Julie's resistance to implementing discourse structures, our conversation still provided her with an opportunity to realize a conflict between her pedagogical responsibilities to support inclusive groupwork and authentically engage students. Through our conversation, Julie found new cognitive salience for reconsidering aspects of her instructional design.

Fermat VFF 5: Lesson Design and Status Issues

In Year 2, we returned to Julie's AP Statistics class for Fermat VFF 5. Julie continued thinking about status issues, especially because her course drew students with very different mathematical backgrounds.

Overview of Lesson

Julie launched a lesson on hypothesis testing with a card game. She held up a deck of cards and told students that if they drew a black card, they would get extra credit on the next test – but if they drew a red card, they would stay after school and sweep her classroom floor. Students eagerly volunteered to draw cards, but each one drew a red card. Julie was jubilant with each draw, oozing with excitement for her soon-to-be spotless linoleum. The students also reacted dramatically, saying, "Hey!" "Oh!" "Wait?" "It's rigged!" with increasing suspicion. After six students drew red cards, Julie told groups work together to generate two hypotheses for what had just transpired.

While groups worked, Julie noticed one student calculating the probability of the sequence of events. Drawing on that student's thinking, she told the rest of the class to calculate the probability, as well. After about a minute, she showed the class that the likelihood of drawing six red cards from a fair deck was $(0.5)^6$ – or 1.5%. She then had students pick a hypothesis based on the probability they calculated. While the main task for the lesson was about a different hypothesis testing context, most of the debrief centered on this card game.

Start of the Debrief

Brette and Lani facilitated the VFF debrief with Julie and Lizette. At the start of the discussion, Julie told us that she had selected focal groups that she felt "had status issues," signaling her continued curiosity about group dynamics. As she elaborated, "I thought maybe from the vibe I got from them through-out the year that there might be some people dominating the conversation." She told us that she had designed her lesson with status issues in mind, using the opening exercise as an example. "There was no math involved there. Maybe a little bit at the end, but every one of them could have contributed to that conversation."

As we explained the lesson to Lizette, Brette said that the point where Julie asked the groups to discuss the likelihood of six red cards in a row "brought out a lot of really interesting conversations in the groups, especially around this status thing, because there were a couple of kids who saw it immediately" and others who were still thinking.

Clips 1 and 2: Looking at Groups' Sensemaking

To review the moment when groups began calculating the probability of drawing six red cards, we watched two clips. In Clip 1, students collaborated very little, but one of them came up with a correct calculation. After watch-ing this, Julie offered an interpretation that most students only understood this calculation procedurally, and that it was hard for them to remember. She summarized her diagnosis as, "This concept, the multiplication rule for inde-pendent events, which we've already taught in regular stats, we go over that so many times. It's one of the hardest things for kids to just remember."

In Clip 2, a different group tried to find the probability of drawing six red cards, but they did not have time to find an answer. They began by discuss-ing how a deck of cards is organized (e.g., "How may red? 26? Out of how many total?"). This group was clearly on the right track – they were making sense of the context and working toward an answer – but Julie interrupted the groupwork before they arrived at a solution. This surprised Julie, who expected students to readily recall that the probability of drawing one red card is 50%, considering how often she used card contexts in her statistics lessons.

Revising the Diagnosis

Brette used Clip 2 to challenge Julie's original interpretation that calculating the probability was difficult because students did not recall the procedure. Instead, Brette proposed two other potential barriers to students' engagement: (1) Julie gave too little time for students to confer, and (2) the contextual underpinnings of the card game were more abstract than Julie expected. Exploring these proposals, Lani asked Julie what understandings she wanted students to develop in the card game. Julie said:

> The big thing I wanted them to leave with was, there are two compet-ing explanations for a result happening. Based on the probability of that result — the likelihood of that result happening given the claim — that should give you evidence of whether you should support one [hypoth-esis] or another.

In other words, Julie's learning goals for her students centered on them con-necting the likelihood of an outcome to the hypothesis.

Brette and Lani continued to press Julie about students' understanding of the deck of cards:

BRETTE: Not all kids have that knowledge super ingrained. It seems like —
JULIE: We play with cards *a lot*. I give them cards for seating charts, every unit. And it's on the table,[1] like it's half and half. You know what I mean? I guess I didn't think that was going to be something they needed to think about.
LANI: Right. We were talking about yesterday, when we were thinking about all of the math of this, we were thinking about how cards — on the one hand, you can make up suits and think about colors. The color is just 50-50, but you have to have sort of an abstract understanding to see —
JULIE: — And we *do* in the probability unit. I let them look at a deck of cards and we come up with a summary. There's 13 of each suit. There's four of each number card. There's four of each face card.
LANI: They did 13 times 4 really quick.
JULIE: They knew the numbers. That's interesting. I would never have thought that was something that group struggled with.

Hearing the details of what her students found challenging versus what came easily surprised Julie. She had assumed that they would quickly recognize the probability of drawing a red card, but many did not. And since Julie did not expect students to have to reason through the number of red cards and black cards in a deck to come to that conclusion, she did not give them enough time to do so.

Eliciting Competing Claims

We then considered Julie's goal of eliciting competing claims. The focal groups collectively had come up with four different explanations for the card game: (1) All the cards are red; (2) students drew cards from the same location in the deck; (3) the cards weren't shuffled; and (4) six students drew red cards by chance or bad luck. Importantly, not all the groups generated *mutually exclusive* explanations, which are necessary for hypothesis testing. For example, one group said that their two explanations were that *all the cards were red* or that *all the students drew from the same place in the deck,* which are not mutually exclusive.

We compared this to Julie's directions to discuss, "What are two possible explanations for how we pulled six red cards in a row?" Lani noted that each group generated two explanations, but that the explanations were not always statistical hypotheses:

LANI: I don't think that they – I mean part of what I think makes teaching really hard is that, once you understand what hypothesis testing is, you're like, "We're looking for two mutually exclusive outcomes." If you're just like, "What are two explanations? Well, here's an explanation, here's an explanation." They didn't get that they needed to be distinct and not both possible simultaneously.

JULIE: Wow. I didn't even think of that.

BRETTE: They definitely didn't all come to the same two things.

JULIE: Right, because there *are* multiple explanations technically.

Julie had to grapple with the precision of her learning goal compared to the loose presentation of what an explanation meant in the launch, another issue of instructional design.

Revisiting Status Issues

Clearly, there were multiple sources for student confusion in the card game. Lizette connected this observation about students' confusion to Julie's concerns about status:

I think that's where maybe when you were talking about the difference in status level, like math understanding-wise – previous math experiences I guess – sometimes, the kids who *haven't* gone through all these honors and AP classes, they have a better sense of the gut-check thing

> and being able to throw out an estimate and a prediction, because they're not used to everything having to be perfectly right all the time.

In other words, Lizette concurred with Julie's initial sense that the card game could have led to productive conversations among students with different prior math experiences, a central concern to Julie. Importantly, Lizette suggested that students who went through high-track classes would be at a disadvantage, as they might be less inclined to use an intuitive "gut-check" because they were socialized to seek correct answers. This offered a more nuanced perspective on status, as the typically low-status students – who "haven't gone through all these honors and AP classes" – would more readily make sense of the card game based on their intuition.

Yet, in listening to groups discuss the probability of drawing six red cards, it seemed that Julie's launch had the opposite effect. Many groups started to figure out the probability but needed more time than Julie had allotted. A few students calculated the probability quickly and gave the answer to their groupmates; they did not have time to share their thought process. Ultimately, asking students to calculate the probability seemed to reinforce status issues by emphasizing speed and shutting down opportunities for discussion and sensemaking.

Lani asked Julie if she saw connections across "the status problem stuff" and "the pacing stuff." She said she did. As we considered why the intuitive question about generating explanations for the six-card streak turned into a status-reinforcing interaction, we arrived at the interpretation that Julie's prompts to find "the two explanations" and "calculate the probability" framed the groupwork around *answer-finding* rather than *answer-generating*. In other words, a more open-ended prompt – something like, "What might be going on here?" or "Why was this surprising?" – could have invited wider participation than a question that could be answered swiftly with a calculator.

Summary of Julie's Learning in Fermat VFF 5

In Fermat VFF 5, the video clips and discussion laid bare the connections across lesson pacing, students' background knowledge, Julie's prompts, and where she expected students to encounter difficulty. During the card came, Julie gave students very little time to generate hypotheses and calculate probabilities. This limited opportunities for more exploratory talk, which many students needed to make sense of the problem context. Additionally, Julie's prompts oriented students toward answer-getting, further shutting down their collaboration.

What Julie initially diagnosed as a problem with students' difficulty recalling a mathematical procedure turned into a more nuanced diagnosis involving many facets of her instructional design. Although the new diagnosis was somewhat overwhelming, it also offered actionable revisions for Julie's future instruction. By attending to her pedagogical actions – especially around pacing and students' background knowledge – Julie could revise her lesson in ways that minimized status differences among students, offering them more ways to engage in mathematical ideas.

Member Check Visit

During our Member Check visit in Year 3, Julie taught the same lesson we observed in Fermat VFF 5, but with clear changes in her instructional design.

Overview of Lesson

Julie launched the lesson with the same card game. As in the previous year, students drew six red cards in a row. The game elicited the same excitement and suspense that we had observed in Year 2. Before sending them to mathematize what they had witnessed, Julie gave students similar prompts, asking them to: (1) Find two possible explanations for the run of red cards; (2) calculate the probability of drawing six red cards in a row; and (3) select a hypothesis based on the probability they calculated.

Despite these similarities, there were notable differences in Julie's lesson design. First, she changed the pacing of her questions. In Year 3, she spent about four additional minutes on the card game, with more time allotted to generating possible explanations and selecting a hypothesis (see Table 8.4).

Clarifying Possible Explanations

In Year 2, we noted students' confusion about generating hypotheses, which was one of the goals of the card game. Specifically, students proposed explanations for the six-card streak that were not mutually exclusive. In Year 3, Julie explicitly invited students to generate multiple explanations, reserving the issue of mutual exclusivity for later in the activity: "Let's just see the likelihood of this happening. But first [...] I need two explanations for how that could have happened. And you can think of whatever explanation you can think of for that."

TABLE 8.4 Comparison of the pacing in Julie's Year 2 and Year 3 card game activities.

	Fermat VFF 5, Year 2	*Member Check, Year 3*
Possible Explanations	1'52"	3'32"
Calculating Probability	3'42"	3'40"
Selecting a Hypothesis	2'11"	4'36"
Total	7'45"	11'48"

In the subsequent whole-class discussion, Julie elicited explanations by asking, "Let's collect as many explanations as we can for how this could have happened." The students came up with ideas like, *the deck was all red, all the red cards were in the middle of the deck*, and *we got unlucky*. To underscore the importance of hypotheses being mutually exclusive, Julie focused on the meaning of a rigged deck. She said to the class:

> So, Oscar says all the cards are red. The deck is not actually half and half. Because you can argue that if I put in *more* red cards than black, then that's also considered rigged, right? So, I'm just going to say the deck is not actually half and half [*writes on the board*] – the standard deck that you were expecting.

Julie consolidated explanations that relied on a rigged deck – whether the distribution of cards was different or the cards were arranged in a certain way – into one hypothesis. This set up a clear contrast with a mutually exclusive hypothesis: "We just got really unlucky." Julie recorded this explanation on the board, as well: "The deck *is* half and half so that really is a normal deck of cards. But this class is super unlucky."

In her facilitation of this discussion, Julie made two other moves to scaffold students' understanding of hypothesis testing. First, she referred to the two explanations as "competing claims," embedding that academic language into the discussion to imbue it with meaning. Second, she asked students which explanation they could find the probability for, supporting students to think about *why* they might do this rather than focusing on *what* the probability might be. This made a more explicit connection between the hypotheses and the probability students would then calculate.

Scaffolding Background Knowledge to Calculate Probability

Before having groups discuss the probability of a six-card streak, Julie asked students, "What's the probability that we get one red?" The class collectively responded with "26 out of 52." Julie replied with "Half-and-half, right? But I'm asking for *six* reds in a row." During this discussion, Julie repeatedly referenced the color composition of a deck of cards, using the phrase "half-and-half." In this way, she provided important background information to support students' understanding of the context.

In contrast to Year 2, Julie did not focus on *how* to calculate the probability. Rather, she narrated *why* they were calculating the probability of drawing six red cards in a row, connecting it to students' intuition:

> The first red [you] drew, you weren't fazed. The second red card, you guys were still like, "Alright." But I did notice as three, four, and five

were drawn, I started to hear the commentary. Did you start to feel like something was up at that point? There was a moment after a certain number of reds were drawn in a row, that you kind of shifted and started to question me. But it didn't happen after the first couple.

Julie then led the class in calculating the probability of drawing one red card (50%), two red cards (25%), and so on, up to six red cards (1.5%) to illustrate the diminishing probability and their growing sense that "something was up."

Using Probability to Select a Hypothesis

Julie reminded students of the importance of 5% – the boundary value of statistical significance – and prompted them to conclude which explanation they believed to be true:

> So, if I were giving you this deck in my hand, there's only a 1.5% chance of six reds in a row. […] I want you to just have a final conversation [with your group] about which of these two explanations you think is true, and I want you to use this probability you found in your explanation.

After a short period of small-group discussion, Julie called the class back together and had students vote for one of the competing claims. She summarized the conclusion, saying, "Usually, you're not going to actually know which explanation is right, but if you have a low enough probability – like there's only a 1.5% chance of this actually occurring – it's not believable." Then Julie revealed her all-red deck of cards.

Evidence of Julie's Learning

In the Fermat VFF 5 debrief, Julie described her original goal for the activity as discussing two competing claims. In her words, the story was "deciding between two possibilities and how you can use a probability to make that decision." However, through the debrief, Julie came to see how her emphasis on the calculation impeded students' understanding of this story and exacerbated status issues. While she used the same card game in Year 3, Julie redesigned key parts of the activity to clarify the "story" of hypothesis testing.

In each whole-class segment, she illustrated how the mathematics contributed to a broader narrative about the likelihood of drawing six red cards. When Julie asked for possible explanations of the six-card streak, she elicited many different ideas. This allowed her to draw on students' intuition about the unfair game and to validate contributions from more students than she had in Year 2. Moreover, by consolidating various explanations of how the deck might be rigged into one hypothesis, Julie was able to make a clearer distinction between two mutually exclusive statements.

Julie further supported this story by explicitly connecting students' intuitions to probability. Rather than relying on students' background knowledge, Julie repeatedly referred to the color composition of decks of cards as "half-and-half." She also emphasized the diminishing probability as students drew one red card after another; this built on their growing sense that something fishy was going on. These changes allowed students to spend less time thinking about *how* to calculate the probability and more time focusing on its *meaning* in relation to what they had just experienced. With that idea clarified, they could better understand what the probability suggested about the likelihood of the competing claims.

Ultimately, Julie's revisions in Year 3 allowed students to make sense of the context – and the probability behind it – regardless of their speed or background knowledge. Her new pedagogical actions reconciled some of the tensions between her pedagogical responsibilities. By spending slightly more time on the card game activity, Julie was able to minimize status differences and provide more opportunities for students to engage in deeper conceptual thinking.

Tensions in Pedagogical Judgment: Connecting Instructional Design to Students' Mathematical Engagement Amidst Competing Pedagogical Responsibilities

Julie's competing pedagogical responsibilities produced very real tensions in her practice and, in turn, for her learning. Specifically, she felt a deep commitment to maintaining authenticity with her students and, through her relationships with them, she sought to broaden access to rich mathematical ideas. In her AP classes, which involved a high-stakes, end-of-course exam, she felt compelled to keep pace with the curriculum and make sure students had a chance to learn the material. During the Member Check interview in Year 3, Julie said that she still felt a sense of "competing goals" and that she continued to reflect on and negotiate these tensions:

> I think the challenges are always the same. They want to memorize sentence frames instead of just talking it through and developing the understanding. And that's hard, because some students really need, like, some frame to plug into.

Despite the challenge, she was committed to learning and growing in her practice.

Julie engaged with these various tensions as she investigated her teaching with us. The VFFs uncovered details of her instructional designs and how students interpreted and made use of them. In Fermat VFF 1, she noticed a common confusion between categorical and numerical data and how unstructured groupwork could inadvertently perpetuate status problems. Instead of simply accepting the difficulties students had in understanding different data types and working together, the VFF inquiry offered a representation of practice that supported a new interpretation: Some of these difficulties were being reinforced by the lesson design, which she could revise to better support students' learning.

In Fermat VFF 5, Julie saw how her pacing and emphasis on calculations undermined the potential inclusiveness of her launch. Reviewing these details once again shifted her initial interpretation of lesson trouble to something more actionable: Instead of diagnosing the difficulty as arising from students' struggling to recall a mathematical procedure, she linked it to aspects of her lesson design that inadvertently exacerbated status dynamics.

Julie reconsidered the relationship between her pedagogical actions and student engagement in many ways. In the Member Check interview, she said that she had started using more deliberate discourse structures to foster students' collaborative talk, particularly with her younger students:

> My 9th-graders, I feel like I put a lot more energy into those types of strategies. I'm not sure why that is, but yeah, [with older students] it ends up being [...], "Oh, they're mature enough to be able to work through it."

In her statistics classes (with primarily 11th- and 12th-graders), she sometimes used pairwork, with the expectation that spending some time collaborating in pairs – rather than groups of three or four – would mitigate problematic student dynamics, like those in Fermat VFF 1. Julie felt the status issues in her AP Statistics classes had improved significantly because these changes to her instructional design allowed her to attend to status issues from the beginning of the year. These examples illustrated Julie's reconceptualization of her pedagogical actions to better align with her pedagogical responsibility to support equitable participation.

Through the VFFs, Julie reflected on students' engagement in light of her instructional design. With rich representations of practice – particularly of groups' discussions – Julie was able to reconceptualize problems related to student learning and inequitable participation as issues with pacing, questioning, or task cues. Fundamentally, this is a more actionable framing of student status: Julie did not have to accept unbalanced group dynamics as inevitable or impose artificial (to her) discourse structures. Instead, she changed her pedagogical actions to reduce barriers to participation, thereby mitigating status differentials and ultimately better reflecting her pedagogical responsibilities.

Discussion: Brad's and Julie's Learning About Teaching Over Time

In these last four chapters, we have offered different perspectives on what and how teachers learned through collaborative sensemaking in the VFFs. This chapter, in particular, focused on teachers' learning about instruction over time, which enabled us to extend our lens beyond specific moments of insight (Chapter 7) to the iterative and often non-linear process of incorporating new discourse practices and instructional designs, experimenting with them in the classroom, getting feedback, and leveraging them in more productive ways.

Both Brad's and Julie's longer learning trajectories stemmed from their strong commitments to their own learning and drew on the resources of their supportive teams. These conditions facilitated their trust in the VFF process and their willingness to be vulnerable with the debrief teams. Interestingly, for both teachers, their new understandings of practice (facilitating groupwork and the relationship between student engagement and instructional design) resulted in them reporting that they learned to listen to their students more closely. We suspect that listening to students during the VFF review alerted them to the value of students' "offstage" talk as a resource in the classroom for understanding how they make sense of ideas and what they experience in lessons. Toward their later VFFs, both teachers used the lesson records to better understand students' responses to their modified practices. In Noether VFF 6, Brad wanted to understand how his redirection of students toward asking group questions met his goal of fostering more mathematical talk, asking us to review the video with this question: "What happens when I walk away?" Similarly, for Fermat VFF 5, Julie shared that she had placed the audio recorders with student groups she had perceived as having status issues, seeking out additional information for her sensemaking about that core concern. For both teachers, the information shared about students' talk and experiences supported new diagnoses of what was happening in their lessons. Brad saw that his redirection catalyzed the mathematical talk he had hoped to hear. Julie realized that the negative status dynamics did not emerge through student dominance as much as through aspects of her lesson design. Listening closely to students through the VFFs shifted the teachers' understandings of their practice.

Of course, students do not need to have microphones on their tables for teachers to listen. Both teachers reported incorporating listening practices in their instruction. For Brad, this resulted in him listening more before he entered students' groupwork conversations. For Julie, this drew her attention to where students experienced challenge with an eye on whether it was about important ideas or tangential ones. Despite these similarities, Brad's and Julie's learning trajectories differed in ways that illuminate distinctive pathways for honing pedagogical judgment and integrating concepts about teaching.

The Evolution of Pedagogical Judgment: Different Pathways Toward Aligning Responsibilities and Actions

Recall that Brad initially described his vision of good student conversation as leading to good mathematical reasoning (Noether VFF 1), and that he gradually grew more attuned to issues of equitable participation and students' authorship of mathematical ideas (Noether VFF 6). We see this shift as bringing into alignment his pedagogical responsibilities (encouraging broader access to mathematics) and his pedagogical actions (structuring groupwork to foster student interdependence instead of talking to whichever student called him over), in part by further refining what encouraging broader access involves – how he interacts with student groups. In contrast, Julie's pedagogical responsibility was rooted from the start in concerns about inclusion, but she did not see how her instructional designs could amplify or ameliorate exclusionary dynamics in her heterogeneous classroom. Her learning, then, focused on reconceptualizing student status as a dynamic attribute that depends on details of her instructional design, not a strictly static property of individual students. For this reason, issues of scaffolding, pacing, and the particulars of her prompts became salient to her as she understood how they contributed to the dominance she sought to interrupt in her classroom.

While Julie's primary pedagogical responsibility was around developing inclusive instruction, she experienced tensions around the institutionally rooted responsibilities of teaching an AP course, with its predetermined curriculum and external exam. The course structure, beyond her immediate control, compelled her to go "really freaking fast," which worked against her inclusion goals. This resulted in onto-epistemic tensions around questions of good teaching. For example, she felt tensions around questions like, *Is it worth structuring student conversations in potentially coercive ways to disrupt status issues?* along with the perennial teaching dilemma, *Is it better to follow the students or the curriculum?* Since Julie continued to experience these tensions, her pedagogical judgment was refined, but not necessarily settled. In comparison, Brad did not report tensions in his learning, so he incorporated new pedagogical actions that supported his commitments, eventually becoming fluent with them.

Concept Integration: Considering and Linking Multiple Facets of Teaching

Through their learning trajectories, both Brad and Julie integrated concepts for teaching. For Brad, his revised concept of groupwork monitoring moved away from a management focus, with the goal of helping students complete their work, to an inclusive focus, with the goal of fostering productive and equitable mathematical conversations. The revised concept shifted his attention to (1) developing a collaborative classroom culture through daily groupwork;

(2) using groupwork monitoring routines that directed students to each other as resources for mathematical sensemaking; and (3) accounting for students' previous discussions in his groupwork interventions. This new concept, with its narrativized understandings of relationships between his actions and his commitments, tied more facets of his instruction together, accounting for his facilitation, students' mathematical engagement, and issues of inclusion.

Julie revised her concept of status issues to move away from a static notion based in students' prior mathematical histories to a dynamic one she could influence through her instructional designs and her moment-to-moment interactional decisions. This new concept offered actionable choices for her as a teacher, integrating her instructional designs, classroom interactions, and students' mathematical engagement. This new concept also required new practices, such as ensuring students had adequate scaffolds to make sense of tasks, as well as increased attention to discursive structures that facilitated her interpretations of students' thinking in the moment. She used group-work and pairwork more intentionally and thought about her discussion prompts and pacing in relation to her learning goals, recognizing the extent to which these instructional design choices could exacerbate or minimize status dynamics.

VFFs and Teacher Learning

Although the two cases presented here differ in many ways, they both high-light the importance of a situative perspective on teacher learning, as well as the relevance of several of our design conjectures in support of that learning. As Brad's and Julie's cases illustrate, the VFFs honored the teachers' commit-ments alongside the research team's observations and concerns. Both Brad's and Julie's learning trajectories illustrate the importance for professional devel-opment to address teachers' existing concepts about and practices for teaching (Conjecture 1) and connect learning activities with teachers' personal goals (Conjecture 2). Furthermore, the discussion of Brad's and Julie's different trajectories toward aligning their pedagogical responsibilities and actions underscores the need to provide adequate and timely feedback on teachers' ongoing efforts to improve their practice, since seeing students' responses to their instruction shifted key understandings for both teachers (Conjecture 5). Both teachers agreed that their learning was enabled by the co-inquiry pro-cess, with its timely feedback about their instructional practice and alignment with their learning goals.

Note

1 Julie referred to the cards that she used to number students' seats – each table had the same card number, with two red cards and two black cards.

References

Ehrenfeld, N., & Horn, I. S. (2020). Initiation-entry-focus-exit and participation: A framework for understanding teacher groupwork monitoring routines. *Educational Studies in Mathematics,* 103, 251–272.

Horn, I. S. (2012). *Strength in Numbers: Collaborative Learning in Secondary Mathematics.* Reston, VA: National Council of Teachers of Mathematics.

Ray-Riek, M. (2013). *Powerful problem solving: Activities for sense making with the mathematical practices.* Portsmouth, NH: Heinemann.

Self, E. A., & Stengel, B. S. (2020). *Toward anti-oppressive teaching: Designing and using simulated encounters.* Boston, MA: Harvard Education Press.

Sengupta-Irving, T. (2021). Positioning and positioned apart: Mathematics learning as becoming undesirable. *Anthropology & Education Quarterly*, 52(2), 187–208.

9

LEARNING ABOUT TEACHER LEARNING

A Situative Theory for Teacher Learning

In this study, we developed a theory of teacher learning by looking at experienced urban secondary teachers' learning about ambitious and equitable mathematics instruction. We partnered with a professional development organization (PDO) that provided over 100 hours of professional development annually, augmenting these resources with a video-formative feedback (VFF) intervention to uncover and trace teachers' learning.

Our theory is necessarily situative. We framed our synthesis of sociocultural studies of teacher learning and studies of teachers' work using Jean Lave's theory of learning theories (1996), arguing that teacher learning can be productively conceptualized as a process of conceptual and cultural change. Lave posits that any theory of learning makes claims about subject-world relations, telos, and mechanism. By viewing teacher learning as a sociocultural process of concept development – the changes in narrativized actions and understandings about teaching – our perspective emphasizes teachers' agency and sensemaking, while also accounting for influence of the cultures of mathematics education and schooling writ large.

In Chapter 1, we proposed that teacher learners' subject-world relations – the way they encounter and use knowledge in, of, and for teaching – is characterized by two forces: their conceptual agency over their practice and its simultaneous embeddedness in the contexts of their work. Consequently, concepts and related teaching practices are shaped by this duality. Additionally, in contrast to well-studied domains of learning such as mathematics, chess, or reading, teaching knowledge is less determinate because of this social embeddedness; it is further complicated by being inherently ambiguous and frequently contested. We summarized these subject-world relations in a set of premises and corollaries in Table 1.1.

DOI: 10.4324/9781003182214-11

Our attention to subject-world relations highlights two important issues. First, embracing the uncertainty of knowledge in teaching – moving away from technocratic conceptions like "best practices" – enables teachers' learning. In rejecting binary and global notions of "goodness," teaching becomes an object of inquiry, a perspectival shift that other researchers have noted in their work. By acknowledging that this binary is ubiquitous in discourses about teaching, we seek to normalize this shift while noting its significance for teachers learning to investigate and more effectively learn from their practice.

In Chapter 2, we described the *telos* of teachers' learning of ambitious and equitable mathematics instruction, the onto-epistemic shifts they make as they learn this form of practice. Making these shifts requires conceptual change around topics like how to organize instructional activities, what math class sounds like, and who belongs in math class, which we captured in dichotomous charts. Importantly, because of their social embeddedness, concepts for teaching do not wholly reside within individual teachers. Instead, they extend across the structures and rituals of schooling and other conceptual resources in teachers' environments, giving meaning to core aspects of practice.

Our description of the telos for responsive teaching extends beyond the details of conceptual change into the details of how concepts themselves are used in teachers' sensemaking. Our review of prior literature suggests that, in addition to the onto-epistemic shifts that capture notions of what things are and how teachers know them, teachers who cultivate responsive forms of practice also develop integrated, ecological understandings, where they consider interconnections across ideas about and practices for core ideas like *the organization of instructional activities* and *who belongs in math class*.

A concept development perspective on teacher learning was crucial to both our design and facilitation of VFFs. It invited us to start with the sense teachers were making of their instruction, rather than starting with an assessment of how well their practices aligned with our pre-determined notions. In many ways, a concept development perspective is critical for moving away from a binary view and toward an expansive view of good teaching, with the latter making teaching discussable and something to explore. If teacher educators and instructional leaders focus primarily on how well teachers conform to pre-determined goals, they essentially evaluate teachers as being either "good" or "bad." Instead, when teacher educators and instructional leaders develop pedagogical concepts by starting with teachers' meanings, they can take a more interpretive stance and help teachers negotiate the multiple (sometimes competing) goals they have for instruction.

Adding to this complexity, as our description of trajectories suggests, teachers' own experiences, commitments, and resources lead different teachers to learn the "same" idea in vastly different ways. A teacher may add a discourse structure to her repertoire of practice because it addresses a keenly felt dilemma, as Doha Arzoomanian did with the huddle (Chapter 7). Another teacher may take a longer time to understand the need for a discourse structure, only incorporating it once its use is apparent through the exploration of a problem of practice, such as Brad Miller's use of the "group question" (Chapter 8). Some concepts for teaching are more diffuse, requiring continual inquiry, as was the case for Julie Woodman's exploration of the relationships between instructional design and students' mathematical engagement, a concept that became more integrated and dynamic over multiple VFF cycles (Chapter 8).

In the last of our theoretical chapters, Chapter 3, we proposed a mechanism for teachers' conceptual change. In our review of studies of teacher learning, we found that intervention design is often conflated with mechanism, making this a frequently overlooked aspect of teacher learning. This omission obscures why particular interventions work, instead offering descriptions of activities (e.g., lesson study, video clubs, instructional coaching) that may or may not work as effectively when the conditions for learning and the participants change. Important to our theoretical aims, our proposed learning mechanism does not just describe teacher learning toward ambitious and equitable instruction, but it also accounts for learning that heads in other directions, thus disentangling mechanism from telos *and* from particular activities.

We describe teacher learning as taking place through the evolution and refinement of pedagogical judgment. As summarized in Figure 3.1, this involves a dialectic between teachers' pedagogical action and pedagogical responsibility, generally (though not always explicitly) through pedagogical reasoning. Pedagogical responsibility reflects teachers' core commitments and is therefore important to their teacher identities. As teachers reflect on the consequences of their pedagogical actions in relation to their responsibilities (pedagogical reasoning), they may realize that their actions do not reflect what they care most about. In that case, they adjust their actions or their responsibilities to keep them sufficiently aligned. When teachers cannot find ways to match their actions to their responsibilities, this can lead to frustration, disengagement, or burnout (Santoro, 2011).

A critical complexity arises when we extricate the development of pedagogical judgment from teachers' pedagogical concepts. Although they are analytically distinct, teachers' concepts are bundled up with their commitments – their pedagogical responsibility – so there is a strong, recursive relationship between concepts and learning. For instance, a teacher committed to modifying instruction to support multilingual learners draws on their concept that *multilingual learners need modifications for support*. The nature of those modifications, in turn, relies on their onto-epistemic stances about what learning is, what mathematics is, and so on – yet regardless of these details, their commitment to support students shapes their pedagogical actions, and, as they reason about the efficacy and utility of these actions, how they refine them. This tight coupling between concepts, onto-epistemics, and learning further explains the finding that teacher learning is an accumulated advantage phenomenon, with teachers who are adept at ambitious and equitable instruction better positioned to productively learn about practice. Like an ouroboros, teacher learning feeds on itself: The onto-epistemics of ambitious and equitable instruction, along with ecological reasoning about the interplay of mathematics, student learning, and teaching, beget further learning and refinement of these practices. Simply put, it is easier to continue to learn about ambitious and equitable instruction once you learn how to teach that way.

Designing for Teacher Learning

In Part 2, we shifted to engage our research question: *How can we use formative feedback to enhance teachers' learning of ambitious and equitable mathematics instruction in urban secondary schools?* In Chapter 4, we described how we used this situative theory of teacher learning to design our intervention. We contextualized our design by critiquing research on professional development, noting that these studies often underconceptualize teacher learning, instead focusing on features that have been shown to correlate with positive outcomes. We also presented our rationale for studying teacher learning of ambitious and equitable instruction with our PDO partners: (1) Our phenomenon of interest would be readily visible; (2) many features of high-quality professional development were already in place; and (3) participants' fellowships limited turnover, facilitating our study of learning over time.

The significance of our best-case participant selection logic – well-supported, highly committed, experienced secondary math teachers in urban schools who belonged to a strong professional network – cannot be overstated. The myriad ongoing challenges that SIGMa teachers encountered in implementing ambitious and equitable mathematics instruction suggests that this form of teaching likely *requires* ongoing support for learning. In particular, as our problems of practice analysis revealed (Chapter 5), many of the challenges that surfaced were related to interactive aspects of practice, such as leading whole class discussions, facilitating groupwork, scaffolding and responding to student thinking, and designing for meaningful engagement.

If the field of mathematics education wants to take equity and inclusion seriously, this finding warrants pause. Classroom marginalization and exclusion often persist *despite* teachers' stated commitments, engagement in professional learning, and good intentions (e.g., Louie, 2017; Esmonde & Langer-Osuna, 2013). SIGMa teachers' pedagogical responsibility centered on building inclusive classrooms, and they were ready, willing, and able to refine their pedagogical actions to align with that goal. That was certainly the case for Brad (Chapter 8). He worked with every bit of feedback we offered about the exclusionary dynamics that his groupwork monitoring practices contributed to, adjusting his practice accordingly. Nonetheless, given his commitments to building an inclusive classroom and his investment in his own professional learning, it is notable that he might not have recognized these dynamics without the feedback of the VFF process.

The ongoing challenge of meaningful inclusion in ambitious and equitable instruction, along with its potential to be met through our intervention, underscores another important point: the significance of teacher commitment for this form of teaching to even be possible. Brad, who was especially receptive to critique, took up our feedback eagerly because of his existing commitments to inclusion; he was not satisfied knowing that his practice did not meet his stated goal of broadening mathematical participation. The PDO's contribution to clarifying and sustaining those values was also important. As an organization and a community, the PDO upheld the cultural model of ambitious and equitable instruction as a feasible and desirable goal, providing teachers with technical assistance, professional development, and, crucially in our analysis, identity resources. Being a "PDO teacher" signaled a set of commitments that made ambitious and equitable mathematics instruction central to teachers' identities and added cognitive salience to our VFF conversations.

Returning to our intervention design, we began with a critique of current work on professional development and its underconceptualization of learning. Typically, summaries of high-quality professional development yield checklists of features, with little attention to how those come together to support teachers' learning. Instead, using a learning environment perspective derived from the *How People Learn* framework (Bransford et al., 1999) and prior research on teacher collaborative talk (Lefstein et al., 2020), we highlighted potentially productive design principles to foster teacher learning. From this exercise, we developed design conjectures to inform our intervention. These were: (1) address teachers' existing concepts about and practices for teaching; (2) align learning activities with teachers' personal goals; (3) draw on knowledge of accomplished teaching; (4) respond to issues that come up in teachers' ongoing instruction; (5) provide adequate and timely feedback on teachers' attempts to improve their instructional practice to support their ongoing efforts; (6) provide a community of like-minded colleagues to learn with and garner support from; and (7) provide teachers with rich images of their own instruction to minimize the burden of recontextualization. In our pilot VFFs, we also developed an eighth design conjecture: Respect teachers' autonomy, agency, and experiences as sensemakers by taking a stance of co-inquiry into practice. Notably, Conjectures 1, 4, 5, 7, and 8 do not correspond to recommendations emerging from prior work on professional development, suggesting that they warrant greater consideration in thinking about the contribution of our study.

Mapping our eight conjectures onto the PDO's current work, we identified where an intervention might extend teachers' learning opportunities. This gap analysis highlighted five of the eight design conjectures, which – perhaps not surprisingly – were the ones not reflected in checklists for high-quality professional development and were unique to the learning environment perspective. These five design conjectures, in turn, shaped our design of the VFF cycle, a video coaching protocol that offered teachers a way to examine their own practice and interpret what was happening in their classroom in a timely fashion relevant to their own learning goals.

Recall that VFF cycles spanned the time before, during, and after focal lessons. Before the lesson, the focal teacher and research team formulated a co-inquiry question. During the lesson, the teacher selected four groups of students to collect audio data on, and the research team recorded the lesson with a whole-class camera and a point-of-view camera, collecting fieldnotes and artifacts. After the lesson, the research team reviewed the recordings with the co-inquiry question in mind. Typically, within 24–72 hours, the researchers met with the focal teacher and their colleagues to debrief the co-inquiry question; the debrief conversations were filmed as a part of our research. We thus had rich documentation of VFFs.

Ultimately, we had 12 participating teachers from five schools, with ten teachers acting as full participants. Over the course of the project, we collected 35 VFF data sets. Thirty-three of the 35 VFF debriefs were coded to capture the problems of practice that emerged in the conversations. We used these data, along with interviews and other observations, to construct learning portraits that captured teachers' evolving understandings about different aspects of ambitious and equitable instruction. In our one-year follow-up visit, nine of the ten full participants showed evidence of using the practices or understandings they had developed through the VFFs.

Revisiting the Design Conjectures

In this section, we revisit the design conjectures that informed our intervention, with an emphasis on the ones that departed from prior research on professional development. We start with the two most complex ones – Conjectures 8 and 7 – and then consider the other conjectures that introduce new ideas about professional development, Conjectures 1, 4, and 5. Additionally, we revisit Conjecture 6 on the importance of teacher community, as the learning mechanisms of this social arrangement warrant continued examination, and we want to explore this widely shared principle through our data.

Conjecture 8: Troubling Co-Inquiry

In Chapters 5–8, we shared learning portraits of our focal teachers, along with our analyses of their problems of practice, to test and refine our theory of teacher learning. Rather than summarizing those chapters here, we instead revisit our design conjectures in light of the findings we reported. To make our argument, we start with the thorniest conjecture, Conjecture 8: *Respect teachers' autonomy, agency, and experiences as sensemakers by taking a stance of co-inquiry into instructional practice and foster interpretive dialogue.*

As we recounted in Chapter 5, we quickly confronted the limitations of our idealized egalitarian stance of co-inquiry. Since our shared investigation centered on uncovering participating teachers' practice – often with closer scrutiny than they had experienced in prior professional development – there was an insurmountably unequal difference in our social risk compared to theirs. The vulnerability of visibility suggests that the egalitarian ideal of our eighth design conjecture requires a significant amendment.

If we were to revise this conjecture, the term *co-inquiry* would come with a set of caveats. First, as we knew, trust is important in this work; however, we underestimated its tenuousness and fragility, given the vulnerability of visibility. Lizette McLoughlin's case (Chapter 6) showed how an unfamiliar person, even in a secondary role, heightened this vulnerability, signaling the need for deep trust, clear communication, and teachers' right of refusal to be observed. Although our human subjects agreements surely would have given Lizette that right, in retrospect, we should have pro-actively informed her of the new researcher's presence, perhaps made time for a longer introduction prior to the observation, and explicitly asked her permission to bring somebody new to her classroom.

Second, any authenticity to our co-inquiry was predicated on sufficiently shared notions of good teaching. Even within our best-case PDO setting, we paid attention to teachers' signals about their readiness to examine different issues. In our video review, we prepared multiple potential discussion points; occasionally, we suggested topics that we felt would be helpful but that the teachers did not want to prioritize. This facilitation choice honored their agency.

However, if we were to apply this design conjecture outside of a context like the PDO – one without shared commitments to ambitious and equitable instruction – we could imagine great frustration on the part of both researchers and teachers about which inquiries to pursue, since these ideas could vary widely. As Anna Sfard (2019) has argued, incommensurable discourses limit the potential for dialogic learning and lead to the prevalent phenomenon of people talking past one another. Even with a shared vision of teaching, we still had to reconcile our different onto-epistemic stances about the nature of good teaching as teachers shifted from evaluative to expansive frames. For this reason, potential VFF partners need to find and cultivate common ground to truly co-inquire into practice. Ultimately, the ideal of co-inquiry warrants healthy skepticism by anybody wanting to take it up, with a particular need to interrogate romantic egalitarian notions and take stock of where vulnerabilities reside.

Finally, in Chapter 6, we described how, as teachers learned to learn in VFFs, some of the inherent vulnerability could be assuaged by pushing back on the baggage of the binary – the idea that teaching is either good or bad, with the video record threatening to make a negative determination. In the case of Greg Kahae, his commitment to and respect for his colleagues helped him participate in VFFs in ways that let go of the binary, with its evaluative discourse, and shift into genuine inquiry. For Veronica Kennedy, she more easily moved towards co-inquiry regarding her colleagues' teaching than her own. For both teachers (and many others), the discourse of the binary (good teaching vs. bad teaching) strongly shaped their engagement in VFFs. Indeed, as Michalinos Zembylas (2005) suggests, such discourses position teachers in relation to their institutions in a manner that exerts social control over their actions and emotions; notions of good teaching linked to institutional conformity – whether that is curriculum coverage, test preparation, or certain teaching approaches – contribute to social reproduction *and* are a source of stress and anxiety when teachers fail to comply.

However, teacher educators and researchers who want to move away from evaluative framings must contend with another contradiction: *We typically have normative views of what good teaching looks like.* There is an inherent dilemma in, on one hand, wanting to invite teachers' curiosity and sensemaking about practice while, on the other hand, holding fast to particular images of how their instruction should develop – the telos described in Chapter 2. Perhaps it is our anthropological training, but we can best describe our management of this dilemma in terms of *emic* and *etic* perspectives on teaching. When we facilitated emically, we made sense of what participants said and withheld judgment, instead pursuing and engaging their meanings; it resulted in fundamentally interpretive dialogue focused on clarifying understandings. When we facilitated etically, we brought in our own ideas and values, often drawing on broader frameworks about teaching to guide teachers' instruction towards their desired goals. This often resulted in questions that probed teachers' problem framing, perhaps suggesting alternative frames. Strategically, we aligned our etic statements to their pedagogical responsibilities – usually by linking the dilemmas and ideas that they brought up to the frameworks and suggestions that we or their colleagues introduced – thereby maintaining our overall alliance with them.

These emic–etic shifts happened within our co-inquiry dialogues, and we think that our sufficiently shared vision of teaching was the only reason that we did not lose participants' buy-in through this process. For example, recall Nadav's prompt to Brad about teacher expectation bias in Noether VFF 6 (Chapter 8). That was a potentially face-threatening issue; however, Nadav had confidence in Brad's commitment to make his classroom more inclusive, and Brad was receptive to the feedback. Similarly, Patty's proposal of the "scaffold" as an alternative to Ezio's "cheat" in Rees VFF 1 built on her knowledge of Ezio's commitments to support students' mathematical agency (Chapter 7). In other words, Conjecture 8 involves the relational, interactional, onto-epistemological, human aspects of this work, and it cannot be overlaid into new situations without significant reflection, investment, and humility on the part of the teacher educators or researchers who attempt to use it.

Conjecture 7: Rich Representations for Teacher Learning

As our discussion about Conjecture 8 underscores, the rich video representations of instruction were a source of vulnerability in our work with SIGMa teachers. At the same time, they were a tremendous source of learning. Conjecture 7 – *provide teachers with rich images of their own classroom teaching to minimize the burden of recontextualization* – held true, even if it came with a double edge. For our teachers, many important learning moments happened because they could see and hear their students interacting without them.

Sometimes, students affirmed what teachers had hoped in their instructional designs – teachers were delighted to hear students talking about mathematics, debating each other's ideas, and building on each other's arguments. Such affirmations matter, since refining practice should account for aspects of teaching that are working as intended. In these instances, teachers were enthused to see how their instructional practices aligned with their pedagogical commitments. The VFFs could be equally valuable when lessons did not go as planned, since the debrief gave them resources to revisit and probe what happened, as Lee Bellver did after struggling to help students understand a word problem (Chapter 7).

Other times, listening to students' interactions, particularly in groupwork, uncovered surprises. We have instances of teachers surprised at students' sophisticated thinking during groupwork – pleasant surprises. Teachers told us that they felt pride in and appreciation for their students in those moments. We have instances of teachers expressing relief that their lesson went better than they recalled, often with a realization that they were being hyper-critical of themselves. But we also have instances of unpleasant surprises, such as when Julie heard groups dismissing mathematically sound ideas in Fermat VFF 1 (Chapter 8). In addition to uncovering new problems of practice, sometimes the unpleasant surprises yielded different diagnoses for problems they already recognized – like when Lizette realized that her groupwork monitoring strategies were not suited for her especially large class (Chapter 6).

Whether teachers' expectations and interpretations were affirmed or challenged, these rich representations grounded our discussions and had greater credibility than we, alone, would have had as observers. The clips also enabled our co-inquiry stance, as we could present classroom moments without coloring them through the language required to describe them, instead inviting collective interpretation as we watched video together. This allowed us to sustain an emic facilitative stance longer than if we had to continually dig into our own observational records: We could invite teachers to elaborate on their perspective and sustain our focus on their sensemaking with questions like, "What did you see or hear that made you think that?" With the help of their colleagues, we could then dig into fine-grained lesson details to investigate their practice.

However, just as the rich representations of instruction bore greater risk to teachers, they posed a risk to students, as well. Again, our best-case sample eased this issue greatly. SIGMa teachers had good rapport with students – the fact that they could meet the district's high bar for consent reflected students' and families' trust in them. But inevitably, our records captured students being anything from merely mischievous to rather crass. If teachers are given access to students' offstage talk, there must be a clear understanding that, outside an extreme situation, students cannot be held responsible for actions that happened when they may have forgotten about the recording device.

Conjecture 1: Addressing Teachers' Existing Ideas and Practices

Many professional learning activities presume a common starting place for teacher learners. They are drastically undifferentiated. With the VFF design, we aimed to fulfill Conjecture 1: *Address teachers' existing concepts about and practices for teaching.* We implemented this in two ways. First, asking teachers what they wanted to inquire about offered a window into what was salient to their learning. Of course, as we described in Chapter 6, for some teachers, the idea of inquiring into teaching was not obvious, and we had to work with them to co-construct inquiry questions. All the teachers caught on eventually, suggesting that this onto-epistemic shift is possible, even within a relatively limited timespan.

Second, through heightened listening (Coles, 2014), we attended to the problems of practice they brought up, as well as their diagnoses of the underlying causes. Returning to the idea of concepts as narrativized actions and understandings, these diagnoses provide small, local theories of cause and effect, like: *Students didn't answer because I asked too many questions in quick succession; students couldn't solve the problem because they forgot the multiplication rule of probability.* Each of these narratives contains within it a set of relations and an implicit sense of causality: Asking too many questions quickly overwhelms students; forgetting a rule makes a problem challenging. By holding up these theories against evidence in the video records, we supported re-interpretations that frequently led to reconceptualizations of practice.

Teachers responded positively to this aspect of our design. As Doha described her VFF experiences in her Exit Interview:

> I feel like I'm really a part of this. It's not like, "Okay, do this, do that, do that right now." Here you feel like 70% is me. The rest, 30%, is wherever I get suggestions from and where I'm going to improve. Before [in other professional development] it was 10% me, 90% is the other person.

Other teachers made similar observations. Furthermore, teachers appreciated the extent to which we probed their pedagogical reasoning and its relationship to their pedagogical responsibility. As Brad said, "You guys [...] just pushing the *whys*, the purpose behind what we're doing."

Conjecture 4: Addressing Issues that Arise

Conjecture 4 was also crucial to our design: *Respond to issues that come up in teachers' ongoing instruction*. Sometimes, these responses alleviated the isolation of teaching when a lesson did not go as planned, helping teachers move past disappointment toward new plans of action. For instance, when Abigail Graham was unsatisfied with students' discourse about factoring quartic functions, the video-rich discussion allowed her to investigate the sources of her dissatisfaction, resulting in new understandings of what had transpired and how she might construct the lesson differently in the future (Chapter 5). Although she and Greg could have discussed their respective experiences with the lesson they co-planned, the VFF activity offered new resources for their reflection.

At other times, VFF debriefs surfaced ongoing issues that came up throughout the curriculum, such as issues of facilitating groupwork or whole-class discussions, as was the case for many focal teachers, including Lizette, Doha, Ezio, Lee, and Brad. Since the teachers' questions shaped our co-inquiry and we used records of their teaching, we could help them make sense of recent events in their classrooms, effectively offering new conceptual infrastructures in their environments.

Our guidelines for facilitating teacher learning opportunities, described in Chapter 5, reflected our goals to help teachers develop concepts and mobilize for future work. However, we often emphasized the former over the latter, digging into clarifying diagnoses of classroom challenges, since our earlier research showed that concept development happens less frequently in schools (Horn et al., 2017) and that conversations such as these strengthen teachers' informal ties in ways that endure beyond the immediate intervention (Horn et al., 2020). Yet we have evidence that mobilizing for future work often took place after VFF debriefs. For instance, in between Noether VFFs 3 and 6, Brad and Marisa co-planned an entire unit that required daily groupwork, incorporating a suggestion from Noether VFF 3 (Chapter 8). Doha not only added the huddle to her repertoire of practice, but she also adapted it in creative ways (Chapter 7). Indeed, the layering of teachers' design work and the feedback of the VFFs worked iteratively – they planned lessons, adjusted their designs on their own, and then saw how things worked on video – giving them lived concepts to support their concept development as we pressed for formal, scientific concepts. In general, as PDO members, SIGMa teachers had many learning resources, and as a result, found many ways to act on their newly diagnosed classroom challenges and newly formed goals.

Conjecture 5: Timeliness and Specificity of Feedback

Our design conjecture about timely and specific feedback departed from summaries of high-quality professional development in the research literature. Specifically, Conjecture 5 stated that *professional learning activities need to provide adequate and timely feedback on teachers' attempts to improve their instructional practice to support their ongoing efforts.*

Importantly, our intervention did not involve frequent visits with SIGMa teachers. As we described in Chapter 4, with most teams, we did three VFF cycles per year, recording some focal teachers only once each year. Nonetheless, the intervention – the closeness to teachers' practice, the high levels of their involvement – was sufficient to support their learning. Revisiting Mary Kennedy's (2019) metasynthesis of professional development research, we see this as aligning with her finding that intensity and duration do not matter as much as supporting teachers' strategic knowledge. In fact, Kennedy argues that time between interventions gives teachers opportunities to integrate what they have learned into their practice before seeking additional feedback, much like Brad and Marisa did with their use of groupwork structures. Timely and specific feedback, with an emphasis on strategy over prescriptiveness, was an important aspect of our design.

Conjecture 6: The Importance of Teacher Community

Although Conjecture 6 did not bring novel ideas to the study or practice of professional development, we revisit it nonetheless through our data. This conjecture stated that *professional learning activities should provide teachers with a community of like-minded colleagues to learn with and garner support from as they work through the challenges inevitable in transformative learning.* As we detailed in Chapter 3, there is substantial evidence for the importance of teacher community in instructional improvement, though it is often misunderstood. The frequent co-occurrence of strong teacher communities and educational settings that successfully reach a broad range of students has led to a widespread misapplication of this research finding, the fallacy that the presence of teacher communities necessarily supports instructional improvement. Sociologists have long suspected that teacher communities enhance professional commitments and professional learning, but only in certain configurations and conditions (e.g., McLaughlin & Talbert, 2001). Deeper links between teacher community and teacher learning are not always clear. We found several instances in our data that offered insights into the role of teacher community in supporting teacher learning.

For Greg, his commitment to and admiration for his colleagues at Noether High School encouraged him to persist with an activity that made him distinctly uncomfortable at first. Coincidentally, two of his colleagues, Abigail and Brad, were the most at ease with the VFF process. Their serious, good-natured engagement with VFFs offered a model for Greg, whose comfort increased over time. In this way, Greg's teacher community offered models of engagement that moderated his individual response to our work.

We had two instances of teachers who experienced challenges in their personal lives which understandably reduced their engagement with Project SIGMa. For Marisa Dawson at Noether High School, this meant she only acted as focal teacher once; for Lee Bellver at Falconer Middle School, this resulted in his absence from several of the team's VFFs. Yet because the intervention involved their teacher communities, Marisa and Lee continued their own learning and contributed meaningfully to their colleagues' learning. For instance, Lee returned from a leave of absence feeling out-of-sorts, as he described in Falconer VFF 10 (Chapter 7), but Doha and Sammie's continuation of the Falconer VFFs allowed him to examine his practice when he was ready. In this way, the teacher community allowed the learning activity to carry on without Lee, with him exiting and re-entering as he was able. In Marisa's case, although she felt limited in her capacity to serve as focal teacher, her participation during the debriefs was important for Brad's learning (Chapter 8). She contributed ideas about supporting student groupwork and helped him carry them through. For both teachers, their communities allowed them to stay engaged with their team's projects, despite the vicissitudes of their personal lives, thus sustaining their learning. We see their stories as examples of legitimate peripheral participation (Lave & Wenger, 1991) that allowed them to stay at the periphery of their communities' practice, moving to the center when they were able; from both positions, they contributed to and benefited from the conceptually rich activity.

Reflections on Experienced Teachers' Learning

When we look across SIGMa teachers' learning, we note some common themes. First, as we have discussed, the onto-epistemic shift of teaching as an uncertain practice lends itself to the co-inquiry that is central to this work. Different participants needed different supports to develop this stance; for some, like Veronica, it was easier to do in relation to their colleagues' teaching than their own.

Second, the practice of critique was crucial (if counternormative) for many participants; as sociological studies have shown, norms of privacy and non-interference prevail at most schools, making critique a risky endeavor (Little, 1982). Nonetheless, as Greg's case illustrates, with enough trust, support, and room for teacher agency, critique can be learned. Critique in teaching would require a shift in most teaching cultures, with its concomitant onto-epistemic shift about how to approach questions of goodness in teaching.

Finally, refining teaching concepts to better align with the commitments of ambitious and equitable mathematics instruction resulted in refinements of teachers' pedagogical judgment. To unpack this last observation, we revisit the subject-world relation premises from Chapter 1 – particularly the Interpretation Premise, the idea that teachers' actions are shaped by their interpretations of instruction. Teachers are subjective actors who pick up different lesson details and filter them through their own experiences and sociohistorical identities. VFFs shifted the salience of different aspects of classroom life, resulting in a kind of tuning that sensitized them to different details in ascertaining whether lessons met their goals, rather than simply noting their relative smoothness. As we asked in Chapter 1, if smoothness is the noise and student learning is the signal, how do we re-orient teachers to more helpful forms of feedback that get them closer to student learning? We see Brad's and Julie's new listening practices as examples of such re-orientation. This resulted in a refinement of their pedagogical judgment, as they developed new sensitivities to shape new interpretations and find new paths of action to better align with their goals.

In other words, refining teachers' pedagogical judgment in line with ambitious and equitable instruction fostered generative interpretive lenses for instruction. For instance, during our Member Check visit a year after our intervention, we observed Brad's revised groupwork practices in action. The difference in his groupwork monitoring was notable. He intervened less often than he had previously, instead circulating and listening to groups as they worked. Before entering student conversations, he asked them to summarize what they had been talking about so as not to make assumptions and to foster students' ownership of the mathematics. These practices were not merely about adopting different pedagogical actions. Instead, he was listening for and eliciting different kinds of feedback from students to support more generative interpretations of student learning and guide his subsequent actions. Brad's concept of *supporting student groupwork* shifted as he realized that actions he had previously viewed as supportive did not align with his commitments to inclusivity. In our theoretical language, his refined pedagogical judgment helped him make generative interpretations about his lessons and respond in ways that better reflected his pedagogical responsibility. This shift is similar to what Anne Edwards (2010) found as novice teachers developed responsive forms of teaching: New interpretations of the social world generate and make possible new courses of action. Refined pedagogical judgment, informed by the onto-epistemic commitments of ambitious and equitable instruction, helps teachers manage a core tension in their subject-world relations. Explicitly addressing the embeddedness of concepts amplifies teachers' conceptual agency. Talking to students differently based on their prior achievement, for instance, reifies the significance of school achievement in teachers' interactions. In contrast, eliciting students' thinking prior to intervening offers teachers new ways of listening, disentangling teachers' expectations of students from the institutionally embedded lens of prior achievement.

Developing generative interpretive lenses on classroom life is crucial, because ambitious and equitable mathematics instruction increases uncertainty and ambiguity in teaching, making generative ways to observe and interpret classroom life even more essential. Additionally, because ambitious and equitable mathematics instruction is a conceptual and cultural change project, identifying new courses of action – hopefully ones that, like Brad's, support more inclusive approaches – is crucial to reculturing classrooms and, ultimately, schools.

Generative interpretive lenses reflect a deeper integration of concepts for teaching. As teachers developed concepts towards ambitious and equitable instruction, they reflected the shifts described in Chapter 2, and the nature of their understandings changed. Not only did we see examples of concept integration – such as Julie's exploration of the relationship between instructional design and student engagement – we also saw concepts become more dynamic. For Brad, students' learning could no longer be adequately predicted by his sense of their prior achievement; instead, he needed to listen differently and take stock of their understanding, as he realized that prior achievement may not predict students' in-the-moment sensemaking. Similarly, for Julie, student status could not be entirely anticipated by their academic or social histories; instead, different situations and instructional designs influenced status patterns. These shifts reflected integrated concepts, relating student learning to teacher actions; they also offered more situated understandings of things like *student achievement, student learning, status,* and *instructional design* – dynamic reconceptualizations with more room for teacher agency.

Finally, our participant sampling for Project SIGMa – the 12 experienced, highly committed urban secondary mathematics teachers who belonged to the PDO – offered, in many ways, a best-case glimpse into the possibilities for teacher learning. Nonetheless, the ongoing challenges that *everyone* encountered with the facilitative aspects of ambitious and equitable mathematics instruction warrants pause. As we described in Chapter 2, research has shown that facilitation issues – like inadequate time or overscaffolding the mathematics – can weaken the learning potential of rich tasks. Given the extensive professional development our participants had attended prior to our intervention, this suggests two (not mutually exclusive) things. First, the teachers' professional learning about interactive aspects of practice was not adequately supported in traditional professional development. Instead, classroom-level feedback is necessary to fine-tune and connect different aspects of instruction. Second, learning facilitative aspects of practice never truly ends – there are always new situations, new group dynamics, new combinations of curricular and institutional pressures, like Lizette's enormous 40-student Calculus class. Our findings suggest that, if the field of mathematics education wants to invest in realizing ambitious and equitable mathematics instruction, new forms of professional education are needed – ones that explicitly engage teachers' understandings and include timely feedback through rich representations of classroom practice.

Limitations of this Study

As with all studies, there are limits inherent in our investigation of experienced mathematics teachers' learning of ambitious and equitable instruction. First, our sample of highly committed volunteers had the benefit of making our central phenomenon visible, but it also raises questions about how their learning processes compare to those of other teachers. As we have suggested, developing the trust required for the vulnerability of visibility in VFFs might be hard to foster if partner teachers' views were incommensurate with ambitious and equitable teaching. We suspect some groundwork would be necessary to discuss problems together and agree on some common framings and potential solutions. That crucial phase of learning is beyond the scope of this study.

The theory of teacher learning we propose here suggests that teacher identity plays an important role. Teacher identity relates to the commitments at the heart of pedagogical responsibility which, in turn, both guide and anchor reflections on pedagogical actions. Because of our visible presence within the PDO and our desire to protect participants' identities within that group as much as possible, we did not name or explore many of the sociocultural dimensions of their identities here. Although they were certainly relevant for teachers' learning, such details could overly specify participants and disclose their identities to colleagues. We know other research addresses questions of teacher identity in relation to teachers' development (e.g., Chen, 2020; Daniels & Varghese, 2020; Harmon & Horn, 2021; Philip, 2011), and we hope that other studies extend our framework by delving more deeply into these critical aspects of teacher identity and teacher learning.

Additionally, our intervention focused on teachers' learning about issues that arise during active instruction, with facilitation being the most frequently explored topic. Since PDO teachers participated in approximately 100 hours of high-quality professional development annually – compared to 36 hours of professional development over three years for the typical U.S. secondary mathematics teacher (Banilower et al., 2018) – this suggests that teacher facilitation, an important dimension of ambitious and equitable instruction, is not adequately addressed in even the best professional development. Of course, other aspects of ambitious and equitable instruction were not adequately addressed in our intervention and came up only peripherally in our problems of practice analysis. For instance, classroom assessment sends strong messages about student competence and, depending on its design, maintains or discourages students' engagement. Although teachers occasionally brought up issues of assessment (and Clark Zapatero even referenced a conversation with Grace that prodded his thinking about it), the VFFs, with their focus on classroom interaction, did not surface such issues frequently or in much depth.

We also suspect SIGMa teachers' engagement in PDO activities primed them for aspects of our work. For example, the monthly PDO meetings always included exploratory mathematical learning, which the teachers almost uniformly reported as a favorite activity. In addition to the chance to extend their mathematical knowledge and recall what it felt like to learn novel topics, those sessions also gave everyone an opportunity to see their colleagues struggle and shine. The deep conversations we had about mathematical content, such as Abigail's lesson on quartics and Julie's on hypothesis testing, would not necessarily have happened among a more representative sample of secondary teachers. Our participating teachers were eager to explore and understood the importance of these facets of the content.

Implications

Our theory of teacher learning of responsive pedagogy has numerous implications for teacher education, professional development, mathematics education in particular and educational research more broadly. We will touch on some of the main points.

Teacher Education

This study highlights the importance of teacher candidates' onto-epistemics in shaping the trajectories of their learning – particularly the overarching onto-epistemic stance of teaching as something that can be inquired into. We see this supported by the work of Jessica Thompson and her colleagues (2013), who studied pre-service secondary science teachers. As teacher candidates, participants were introduced to methods for supporting rich classroom discourse. As full-time teachers, they took different trajectories with these practices – sustaining, adapting, or abandoning them – depending on their views of science, learning, and their school communities. Applying our theoretical language to their findings, the teachers' onto-epistemic commitments, local conceptual resources, and identification with school-based teacher communities influenced their trajectories in facilitating rich classroom discourse. Deliberate attention, then, to the onto-epistemics that underlie particular practices is crucial in teacher education, even if candidates sometimes resist the practices for not being "useful" (see the Utility Premise).

Additionally, it is crucial to develop teacher education pedagogies that help novices develop generative interpretive lenses for practice – ones that uncover the hidden curriculum for teacher learning embedded in the conceptual infrastructure of schooling. Accordingly, teacher education should offer an onto-epistemic stance on teaching itself as an uncertain practice. We see one example of such pedagogy in Elizabeth Self and Barbara Stengel's (2020) work on their use of simulated encounters with pre-service teachers. In these encounters, pre-service teachers interact in tricky educational situations where there is no one "right" course of action. The teachers interact with actors playing the role of a student, parent, or colleague in a standardized situation for which the teacher has incomplete information. Based on a pedagogy from medical education, the encounters are filmed and used as objects for individual and collective reflection. In one example, prospective science teachers confer with a student who has disengaged in class after the teacher rejected her proposal for a project on intelligent design. Self and Stengel have developed sophisticated approaches for uncovering the onto-epistemic and teacher identity issues that arise in these encounters. Years afterwards, their former students describe how the simulations informed their pedagogical reasoning, suggesting they are powerful object lessons about their pedagogical actions and responsibilities.

Notably, uncertain onto-epistemics on teaching and a focus on developing generative interpretative lenses run counter to the numerous forces that promote technocratic solutions for teachers. The very language of "best practices" elides the consequence-focused questions of *best for whom?* and *best for what purpose?* A situative perspective on teacher learning, with its insistence that contexts are a part of concepts, instead necessitates responsive practice by attending to teachers' sensemaking processes. Indeed, given the ouroboros of teacher learning, fostering generative interpretive lenses may be crucial to teachers' career-long learning because of the endemic interpretive problems that arise when teachers try new practices with disappointing results: *Did this not go well because I am new at it? Did this not go well because this is not a workable approach? Or did this not go well because it was hailing outside and students were distracted?* These are difficult questions to sort out without productive sources of feedback. Tuning teachers to such feedback – especially de-emphasizing smoothness as the primary mark of a successful lesson – stands to guide their own experimentation when they leave teacher education.

Although our facilitation was not the focus of our study, we suspect that some of the facilitative moves we developed and used – building on teachers' strengths, reflecting, considering alternative courses of action, clarifying values – are already in use by many teacher educators and classroom mentors. (Indeed, this is the context where we ourselves developed these practices.) We are heartened to see increased research attention on the work of facilitation (e.g., Gibbons et al., 2021), and it is our hope that our theoretical framework can contribute to that work, as well.

Professional Development

Since our study was in the context of professional development, we have already noted its relevance. To reiterate a few important points, a learning design (rather than a checklist) approach to planning professional development stands to support teacher learning. This perspective presses professional development leaders to account for teachers' current practices and understandings and then create experiences to support their growth toward desired practices. Importantly, our findings also highlight the importance of fostering cognitive salience for desired practices – the need to engage teachers' pedagogical responsibilities and help them incorporate new practices that align adequately with those commitments. Issues of alignment, productive friction, and incommensurability between particular teachers and practices promoted in professional development matter, and future research should document the work of helping teachers shift their commitments in ways that make the practices of ambitious and equitable instruction salient for their learning.

In outlining the limitations of our study, we also mentioned something it revealed: Abundant, high-quality professional development does not adequately support teachers' development of facilitation practices. We consider some possibilities for this finding. First, because teachers have conceptual agency, this could be, in part, due to some practices' misalignment with their teaching identities. Recall Julie's reluctance to use certain discourse structures: Even though she acknowledged that they might mitigate some status dynamics, she felt that these structures conflicted with her commitment to authenticity. Professional development needs to better surface and engage teachers' commitments and onto-epistemics in teaching. Instead of seeing teachers like Julie as "resistant," sources of hesitation should be discussed and interrogated.

Alternatively, the Asynchronicity of Reflection Premise may be at play, as most professional development takes place apart from active instruction. Teachers might learn facilitation strategies in workshops and they may even rehearse them with their colleagues. However, this approximation of practice (Grossman et al., 2009) may not sufficiently capture the simultaneous and conflicting goals teachers experience during instruction – this is the recontextualization problem of Conjecture 7. Instead, our perspective on learning as the refinement of pedagogical judgment suggests the need to foster a productive dialectical relationship between action and purpose, helping teachers uncover the consequences and trade-offs of various actions. This is, in effect, a shift from technocratic views of teachers as implementers to situative views of teachers as instructional designers (Horn & Gresalfi, 2021).

Mathematics Education

Our finding about facilitation deserves the field's attention. The practices advocated for in documents promoting ambitious and equitable mathematics instruction require skillful facilitation. As Nicole Louie (2017) has compellingly shown, commitments to this form of instruction, coupled with investments in professional learning, do not necessarily overcome the exclusionary culture of mathematics classrooms. In other words, the need for adequate alignment between teachers' pedagogical responsibility and the goals of ambitious and equitable teaching is merely a starting point: It is necessary, but clearly not sufficient, for overcoming both the pull of the conceptual infrastructure of schooling and whatever biases emerge when teachers react quickly, as they must during the flow of instruction – conditions that heighten the pull of habitual responses. If mathematics educators want ambitious and equitable mathematics instruction to become more widespread, the conditions of teaching need to change, because these are also the conditions of teacher learning. Limited preparation time, relative isolation, and little substantive feedback on instruction offer teachers few resources for the kind of transformative learning that this difficult, ambiguous form of instruction requires.

Another common concern for teachers involved questions of scaffolding – adequate supports to facilitate students' productive mathematical conversations. When Marisa listened to students sorting out basic definitions in her Geometry lesson, taking away time from the core issues for the day, Greg suggested a labeled diagram to help them get into the problem more quickly. When Lizette realized her students needed more resources (beyond her direct attention) for groupwork, she developed a practice of using anchor charts. Similarly, we discussed Doha's reconceptualization of when and how to interrupt lessons when students encounter difficulties, Ezio's distinction between "the cheat" and "the scaffold" to help students persevere, and Lee's need for clearer anchors for ambiguous problems. Issues of scaffolding came up frequently in Julie's exploration of relationships between instructional design and student engagement; the revisions to the card activity in our Member Check visit involved better resources for students (as well as better pacing, which of course, is related).

Building adequate scaffolds for rich tasks is a non-trivial – and highly situative – area for pedagogical judgment. Teachers need to account for what their students know about tasks' content and context; they need to anticipate what students will need and how to give it to them without "overscaffolding" and reducing cognitive demand. Moreover, they need to consider the mathematics of the task to identify the "steep parts" to make sure students exert themselves in productive places. In other words, building adequate scaffolds is a situative, interpretive problem, and it matters for teachers' facilitation. Julie's revised card game demonstrated how better resources helps teachers meet their goals for students' mathematical sensemaking. Mathematics educators should further investigate issues of productive scaffolding of student conversations, with a particular emphasis on how teachers learn to identify the need for scaffolding, interpret those instructional moments, and offer supports.

Educational Research

A situative theory of teacher learning has implications for many topics in educational research. Our study highlights the complexity of the say/do problem in teaching – that is, the gap between how people describe their activity and what they actually do. When Lee Shulman (1986) asserted that "the teacher is not only a master of procedure but also of content and rationale, and capable of explaining why something is done" (p. 13), he captured one important issue in accomplished teaching while overestimating another. As Charles Munter and Richard Correnti's (2017) research has shown, articulating a vision of high-quality mathematics instruction does, in fact, enable teacher learning toward ambitious and equitable practice. Explanations, with their underlying causal narratives and onto–epistemics, are an important aspect of pedagogical concepts, and pedagogical responsibilities conveyed in these visions may reflect what is cognitively salient in teachers' personal learning projects. At the same time, Shulman underestimated the limitations of teachers' perception, with its inherent partiality and inevitable biases, which make the ephemera of teaching (e.g., subtleties of pacing or participation patterns) especially underrepresented in teachers' explanations of their practice. In other words, teachers' explanations are important, but they are not everything.

Expanding this observation with Project SIGMa data, consider how we paired fine-grained analyses of instructional practice with recordings of teachers discussing and reflecting on their pedagogical choices. Undoubtedly, if researchers talked with or administered a survey to SIGMa teachers about how they teach, the researchers would be dazzled and delighted. Among the many things PDO teachers learned through professional development, they talked about ambitious and equitable mathematics instruction in impressive ways.[1] We do not think that their sophisticated ability to describe classroom teaching misrepresents their knowledge; instead, we see telling stories about teaching as one form of knowledge – one that likely guides their growth – yet we must recognize what it does and does not reflect about actual instruction. We have shown here some of the crucial details that do not make it into such accounts, such as the finer-grained concepts invoked when teachers diagnose why a lesson did not go as expected. Understanding the ontological status of teachers' explanations in relation to practice has implications for different research methods, including how to interpret teacher self-reports or survey data.

Our situative perspective also underscores the influence of policies, work conditions, and social conditions on teaching practice. In Noether VFF 2, Greg deemed a proposal to integrate technology into a redesigned lesson as unrealistic since it was onerous to check out the school's Chromebooks (Chapter 5). We also showed how both Lizette's and Doha's practices were shaped by their large class sizes of over 40 students, a condition which generated new learning demands for them (Chapters 6 and 7). For Ezio, the school's "hidden tracking" and the surrounding community's gentrification raised new issues for him around students' mathematical confidence. He saw students recognize the hidden tracking, with those in the (unofficial) "low class" describing themselves as "dumb." This made issues of status and inclusion salient for his learning (Chapter 7). As we have said repeatedly, teachers' work conditions are their learning conditions.

Another methodological contribution of this study is centering teacher sensemaking across multiple teacher groups, providing an example of a team ethnographic study to investigate teacher learning. Although team ethnographies have been used to study and compare other aspects of education (e.g., Creese et al., 2008), they are not typically used in the study of teacher learning. This is important, because as Lynn Goldsmith and colleagues' (2014) review of mathematics teacher learning showed, 73% of studies involve small participant samples ($n \leq 5$) or larger samples ($n \geq 21$), with relatively few in the midrange. Our sample size allows us to thickly describe specific teachers, their settings, and their learning, while offering generative comparisons across them. There are numerous findings that we only uncovered through systematic comparison. For example, Abigail's and Brad's comfort with critique contrasted with other teachers' discomfort, leading us to pursue their personal histories of critique. Additionally, Goldsmith and colleagues' review shows that research on teacher learning skews towards elementary teachers, perhaps because they are an easier group for researchers to access. The conditions for secondary teacher learning are inherently different, with more time constraints, increased student contacts, and more sources of institutional pressure to teach specific content. These different work conditions constitute different learning conditions.

But work conditions can be moderated, to some extent, by external organizations and resources. Our study also suggests that the PDO sustained and enhanced teachers' professional commitments, not only by providing resources like technical assistance and professional development, but by fostering a cultural model of responsive practice and enhancing teachers' identification with it, thus increasing its salience in their learning. In the PDO's external evaluator reports, the PDO teachers often attributed their longevity in teaching (at least in the sometimes-challenging work conditions of the district) to the PDO and its resources.

Finally, our study offers insight into the often-observed regression to the norm in teaching, a trend that is especially upsetting for those who hope to make inclusive education a widespread reality. The gravitational pull of institutional alignment in teachers' reasoning and action helps explain Louie's (2017) finding of the persistence of exclusionary cultures in mathematics classrooms. Social reproduction is a function of schools in society, and U.S. schools contribute heavily to social inequity (Domina et al., 2017). As Jeannie Oakes and John Rogers (2007) argue in their essay reflecting on why efforts to equalize education fare so poorly in the U.S.:

> Rather than being at odds with our cultural values, inequality is endemic to the logic of our society and to the role schools play in it. At least three powerful cultural "logics" shape how people make sense of the schooling that society provides to various groups of students: the logic of scarcity; the logic of merit; and the logic of deficit.
>
> *(p. 196)*

It is these logics we are helping teachers resist when they come to disentangle core concepts about teaching from their institutional origins. Since these concepts are distributed across the organization of schooling, they do not require individual teacher investment to continue their enactment (although their persistence is greatly facilitated when teachers are invested in them). Even better than attempting to redress these logics teacher-by-teacher through individual interventions like ours, we should, as a field, seek to vastly reorganize our educational institutions to support the flourishing of all children.

Note

1 To verify our sense of this, one of our early interviews used Munter's (2014) visions of high-quality mathematics instruction protocol and Jackson and colleague's (2017) visions of student mathematical capabilities protocol. Although we did not rigorously code their responses, our shared impression was that their descriptions of mathematics teaching met the criteria as both ambitious and equitable.

References

Banilower, E. R., Smith, P. S., Malzahn, K. A., Plumley, C. L., Gordon, E. M., & Hayes, M. L. (2018). *Report of the 2018 NSSME+*. Chapel Hill, NC: Horizon Research, Inc.

Bransford, J. D., Brown, A. L., & Cocking, R. R. (1999). *How people learn* (Vol. 11). National Academy Press.

Chen, G. A. (2020). "That's obviously really insensitive:" Attuning to marginalization in a parent-teacher encounter. *Cognition and Instruction, 38*(2), 153–178.

Coles, A. (2014). Mathematics teachers learning with video: the role, for the didactician, of a heightened listening. *ZDM, 46*(2), 267–278.

Creese, A., Bhatt, A., Bhojani, N., & Martin, P. (2008). Fieldnotes in team ethnography: Researching complementary schools. *Qualitative Research, 8*(2), 197–215.

Daniels, J. R., & Varghese, M. (2020). Troubling practice: Exploring the relationship between Whiteness and practice-based teacher education in considering a raciolinguicized teacher subjectivity. *Educational Researcher, 49*(1), 56–63.

Domina, T., Penner, A., & Penner, E. (2017). Categorical inequality: Schools as sorting machines. *Annual Review of Sociology, 43*, 311–330.

Edwards, A. (2010). How can Vygotsky and his legacy help us to understand and develop teacher education? In V. Ellis, A. Edwards, & P. Smagorinsky (Eds.), *Cultural-Historical Perspectives on Teacher Education and Development*. Routledge.

Esmonde, I., & Langer-Osuna, J. M. (2013). Power in numbers: Student participation in mathematical discussions in heterogeneous spaces. *Journal for Research in Mathematics Education, 44*(1), 288–315.

Gibbons, L. K., Lewis, R. M., Nieman, H., & Resnick, A. F. (2021). Conceptualizing the work of facilitating practice-embedded teacher learning. *Teaching and Teacher Education, 101*, 103304.

Goldsmith, L. T., Doerr, H. M., & Lewis, C. C. (2014). Mathematics teachers' learning: A conceptual framework and synthesis of research. *Journal of mathematics teacher education, 17*(1), 5–36.

Grossman, P., Compton, C., Igra, D., Ronfeldt, M., Shahan, E., & Williamson, P. (2009). Teaching practice: A cross-professional perspective. *Teachers College Record, 111*(9), 2055–2100.

Harmon, M. D., & Horn, I. S. (2021). Seeking healing through Black Womanhood: Examining the affordances of a counterspace for Black Women pre-service teachers. *AILACTE Journal*, 105–121.

Horn, I. S., Garner, B., Kane, B. D., & Brasel, J. (2017). A taxonomy of instructional learning opportunities in teachers' workgroup conversations. *Journal of Teacher Education, 68*(1), 41–54.

Horn, I., Garner, B., Chen, I. C., & Frank, K. A. (2020). Seeing colleagues as learning resources: The influence of mathematics teacher meetings on advice-seeking social networks. *AERA Open, 6*(2), 2332858420914898.

Horn, I., & Gresalfi, M. (2021). Broadening participation in mathematical inquiry: A problem of instructional design. In C. Chinn, R. Duncan, & S. Goldman (Eds.), *International Handbook of Inquiry and Learning*. Routledge.

Jackson, K., Gibbons, L., & Sharpe, C. (2017). Teachers' views of students' mathematical capabilities: Challenges and possibilities for ambitious reform. *Teachers College Record, 119*(7), 1–43.

Kennedy, M. M. (2019). How we learn about teacher learning. *Review of Research in Education, 43*(1), 138–162.

Lave, J. (1996). Teaching, as learning, in practice. *Mind, Culture, and Activity, 3*(3), 149–164.

Lave, J., & Wenger, E. (1991). *Situated learning: Legitimate peripheral participation*. Cambridge University Press.

Lefstein, A., Louie, N., Segal, A., & Becher, A. (2020). Taking stock of research on teacher collaborative discourse: Theory and method in a nascent field. Teaching and Teacher Education, 88, https://doi.org/10.1016/j.tate.2019.102954

Little, J. W. (1982). Norms of collegiality and experimentation: Workplace conditions of school success. *American Educational Research Journal, 19*(3), 325–340.

Louie, N. L. (2017). The culture of exclusion in mathematics education and its persistence in equity-oriented teaching. *Journal for Research in Mathematics Education, 48*(5), 488–519.

McLaughlin, M. W., & Talbert, J. E. (2001). *Professional communities and the work of high school teaching*. University of Chicago Press.

Munter, C. (2014). Developing visions of high-quality mathematics instruction. *Journal for Research in Mathematics Education, 45*(5), 584–635.

Munter, C., & Correnti, R. (2017). Examining relations between mathematics teachers' instructional vision and knowledge and change in practice. *American Journal of Education, 123*(2), 171–202.

Oakes, J., & Rogers, J. (2007). Radical change through radical means: Learning power. *Journal of Educational Change, 8*(3), 193–206.

Philip, T. M. (2011). An "ideology in pieces" approach to studying change in teachers' sensemaking about race, racism, and racial justice. *Cognition and Instruction, 29*(3), 297–329.

Santoro, D. A. (2011). Good teaching in difficult times: Demoralization in the pursuit of good work. *American Journal of Education, 118*(1), 1–23.

Self, E. A., & Stengel, B. S. (2020). *Toward Anti-Oppressive Teaching: Designing and Using Simulated Encounters*. Harvard Education Press.

Sfard, A. (2019). Learning, discursive faultiness, and dialogic engagement. In N. Mercer, R. Wegerif & L. Major (Eds.), *The Routledge International Handbook on Dialogic Education* (pp. 88–99). Taylor & Francis.

Shulman, L. S. (1986). Those who understand: Knowledge growth in teaching. *Educational Researcher, 15*(2), 4–14.

Thompson, J., Windschitl, M., & Braaten, M. (2013). Developing a theory of ambitious early-career teacher practice. American Educational Research Journal, 50(3), 574–615.

Zembylas, M. (2005). Discursive practices, genealogies, and emotional rules: A poststructuralist view on emotion and identity in teaching. *Teaching and Teacher Education, 21*(8), 935–948.

INDEX

Note: Page references in *italics* indicate figures, and page references in **bold** indicate tables. Names in *italics* are pseudonyms.